Synthesis Lectures on Engineering, Science, and Technology

The focus of this series is general topics, and applications about, and for, engineers and scientists on a wide array of applications, methods and advances. Most titles cover subjects such as professional development, education, and study skills, as well as basic introductory undergraduate material and other topics appropriate for a broader and less technical audience.

Geancarlo Abich · Luciano Ost · Ricardo Reis

Early Soft Error Reliability Assessment of Convolutional Neural Networks Executing on Resource-Constrained IoT Edge Devices

Springer

Geancarlo Abich
University of Santa Cruz do Sul
Santa Cruz do Sul, Rio Grande do Sul, Brazil

Luciano Ost
Loughborough University
Loughborough, UK

Ricardo Reis
Federal University of Rio Grande do Sul
Porto Alegre, Rio Grande do Sul, Brazil

ISSN 2690-0300   ISSN 2690-0327 (electronic)
Synthesis Lectures on Engineering, Science, and Technology
ISBN 978-3-031-18601-1   ISBN 978-3-031-18599-1 (eBook)
https://doi.org/10.1007/978-3-031-18599-1

This Springer imprint is published by the registered company Springer Nature Switzerland AG
The registered company address is: Gewerbestrasse 11, 6330 Cham, Switzerland

Preface

Machine Learning (ML) algorithms have provided straightforward solutions to a wide range of applications. The high computational demand of such algorithms limits their adoption in resource-constrained devices, which typically rely on reduced memory footprint and low-power components (e.g., microcontrollers and processors). While performance improvement, customized, and reduced-precision implementations of ML algorithms have been studied extensively, their susceptibility to soft errors caused by radiation particles is still an open question. In this regard, due to their flexibility and high simulation performance, researchers are using Virtual Platform (VP) frameworks to assess the soft error reliability of complex systems considering several software stack components running on top of commercial processors. While the gain in simulation speed is trivially observed in VP simulators based on Just-In-Time (JIT) dynamic binary translation, the soft error assessment consistency of underlying fault injection frameworks remains unclear. In this regard, the *main contribution* of this Book is to provide, at early design phases, a consistent and extensive soft error reliability assessment of ML algorithms developed with specialized libraries that enable the execution of such applications in resource-constrained Arm processors. The *first goal* of this Book is to analyze the consistency of the soft error reliability assessment of a JIT-based fault injection framework (SOFIA) against fault injection campaigns conducted with event-driven simulators (i.e., more realistic and accurate platforms) considering single-processor architectures. Considering the consistency of the results conducted with SOFIA, the *second goal* of this Book is to early investigate and identify the correlation between fault injection results, NN optimized kernels, and reduced-precision parameters of Convolutional Neural Networks (CNNs) executing on resource-constrained IoT devices. Such a study aims at evaluating the balance between relative performance and reliability to promote the use of software-based mitigation techniques to improve soft error reliability. Understanding that adopted CNNs are vulnerable to soft errors, the *third goal* of this Book is to evaluate the impact of soft errors in the code, parameters, and data stored in the memory units of IoT edge devices considering the optimized libraries and the reduced precision used in such ML models. Besides that, we also developed a parallel CNN version as an attempt to increase performance while evaluating the impact of multi-threaded parallelism in the soft error reliability w.r.t. the original sequential version. In this sense, the results conducted in this Book comprise more than 14.8 million fault injections considering distinct case studies, architectures, number of cores, OSs, and parallelization libraries. The consistency evaluation shows that SOFIA is more than $1000\times$ faster than cycle-accurate simulators while preserving the soft error analysis accuracy (i.e., mismatch below to 10%). The early soft error reliability assessment of CNN executing on resource-constrained IoT Edge devices shows that the occurrence of critical faults varies depending on the instruction set architecture, the layer where the faults are injected, and the precision bitwidth of the convolutional layers. With that in mind, promoting the lightweight Register Allocation Mitigation Technique (RAT) gives the best relative performance, memory utilization, and soft error reliability trade-offs w.r.t. a more traditional replication-based mitigation approach. Furthermore, results from fault injections in memory sections show that stored binaries and trained parameters tend to have more critical faults than register files. Moreover, this Book's contributions enable us to advance our study to different multiprocessor platforms to gain performance while maintaining low energy costs during execution, which implies different reliability parameters to be considered both in execution and in data stored in memory.

Porto Alegre, Brazil   Geancarlo Abich
September 2022   Luciano Ost
                                                                                                                     Ricardo Reis

Acknowledgements

We would like to thank to acknowledge with gratitude, the support of individuals and institutions that have made this publication a reality. First, we would like to give special recognition to research partners and colleagues, in special to Felipe Rosa, Jonas Gava, Rafael Garibotti, Vitor Bandeira, Felipe Bortolon, and Guilherme Medeiros, who contributed significantly to the discussion and development of this work.

Our gratitude also goes to the support offered by the Brazilian National Council for Scientific and Technological Development (CNPq—Brazil), Coordination for the Improvement of Higher Education Personnel (CAPES), Foundation for Research Support of the State of Rio Grande do Sul (FAPERGS). We also want to thank the Graduate Program on Microelectronics (PGMICRO) and Graduate Program on Computer Science (PPGC), Institute of Informatics, Federal University of Rio Grande do Sul (UFRGS).

The authors would like to thank Duncan Graham, Larry Lapides, Simon Davidmann, and Imperas Software Ltd. for their technical support and access to their models and simulator.

We would also like to acknowledge the Arm University Program and the Loughborough University for their support.

We acknowledge all the contributors, without whom this publication would not have been possible.

Thank you everyone.

Geancarlo Abich
Luciano Ost
Ricardo Reis

Contents

Acronyms

ABFT	Algorithm-Based Fault Tolerance
AI	Artificial Intelligence
ANN	Artificial Neural Network
API	Application Programming Interfaces
Arm	Advanced RISC Machines
ASIC	Application-Specific Integrated Circuits
AVF	Architecture Vulnerability Factor
AVX	Advanced Vector Extensions
BITFLIPS	Basic Instrumentation Tool for Fault Localized Injection of Probabilistic SEUs
BNN	Binarized Neural Network
CIFAR	Canadian Institute for Advanced Research
CMix-NN	CMSIS-based Mixed precision API for Neural Networks
CMOS	Complementary Metal-Oxide-Semiconductor
CMR	Conditional Modular Redundancy
CMSIS	Common Microcontroller Software Interface Standard
CMSIS-NN	CMSIS API for Neural Networks
CNN	Convolutional Neural Networks
CPU	Central Processing Unit
CUDA	Compute Unified Device Architecture
DFI	Direct Fault Injection
DNN	Deep Neural Network
DSP	Digital Signal Processor
EAFC	Extrapolated Absolute Failure Count
ECC	Error Correcting Code
EDDI	Error Detection by Duplicated Instructions
FDSOI	Fully Depleted Silicon On Insulator
FI	Fault Injections
FIM	Fault Injection Modules
FinFET	Fin Field Effect Transistor

FINN	Fast Inference for binarized Neural Networks
FIT	Failure-in-Time
FPGA	Field-Programmable Gate Arrays
FPU	Floating-Point Unit
GCC	GNU Compiler Collection
GEMS	General Execution-driven Multiprocessor Simulator
GPU	Graphic Processing Units
HDL	Hardware Description Languages
HPC	High Performance Computers
HWC	Height-Width-Channel
IBM	International Business Machines
ICN	Inter-Channel Normalization
IoT	Internet of Things
IR	Intermediate Representation
ISA	Instruction Set Architectures
JIT	Just-In-Time
KIPS	Kilo Instructions Per Second
kNN	K-Nearest Neighbor
LLVM	Low Level Virtual Machine
M*DEV	Multicore/Multiprocessor Software Development Kit
MAC	Multiply-And-Accumulate
MARSS	MIPS Assembler and Runtime Simulator
MIPS	Million Instructions Per Second
ML	Machine Learning
MMC	Microarchitectural Memory Component
MMU	Memory Management Unit
MNIST	Modified National Institute of Standards and Technology
MOSFET	Metal-Oxide-Semiconductor Field-Effect Transistor
MPI	Message Passing Interface
MWTF	Mean Work To Failure
NLP	Natural Language Processing
NN	Neural Networks
NPB	NAS Parallel Benchmark
nZDC	near Zero silent Data Corruption
OMM	Output Memory Mismatch
ONA	Output Not Affected
OpenMP	Open Multi-Processing
OS	Operating Systems
OVPsim	Open Virtual Platform Simulator
PC	Per-Channel
PL	Per-Layer

P-TMR	Partial Triple Modular Redundancy
QCL	Quantized Convolutional Layer
QEMU	Quick Emulator
RAM	Random Access Memory
RAT	Register Allocation Technique
RECCO	REliable Code COmpiler
ReLU	Rectified Linear activation Unit
ResNet	Residual Network
RNN	Recurrent Neural Networks
RPi2	Raspberry Pi 2
RTL	Register-Transfer Level
RTOS	Real-Time Operating System
SASSIFI	SASSI-based Fault Injector
SBU	Single Bit Upsets
SDC	Silent Data Corruptions
SEB	Single Event Burnout
SEE	Single Event Effects
SEGR	Single Event Gate Rupture
SEL	Single Event Latchup
SET	Single Event Transient
SEU	Single Event Upset
SIMD	Single Instruction Multiple Data
SMLAD	Signed Multiply Accumulate Long Dual
SNN	Spiking Neural Network
SOFIA	Soft error Fault Injection Analysis
SPARC	Scalable Processor Architecture
SVM	Support Vector Machine
SWAR	SIMD Within a Register
SWIFI	Software Implemented Fault Injection
SWIFT	Software Implemented Fault Tolerance
TMR	Triple Modular Redundancy
TPU	Tensor Processing Unit
UART	Universal Asynchronous Receiver Transmitter
UAV	Unmanned Aerial Vehicles
UT	Unexpected Termination
VP	Virtual Platforms
WCET	Worst-Case Execution Time

Introduction

Computing is a reality in human daily lives since computers are used from wearable devices to complex High Performance Computers (HPC). Relevant improvements in internet protocols and the computational efficiency of emerging technologies have made communication between different devices more accessible than before [1]. Recent reports, from communication technology enterprises [2] estimate that there will be around 25–32 billion devices connected to the Internet in the coming years (i.e., $\sim$ 3.6 networked devices *per capita* by 2023). However, the popularity of such systems generated a high demand for data to be processed in a timely manner [3, 4]. In this sense, researchers started to use Machine Learning (ML) models to identify patterns in large-scale datasets, aiming to classify or predict the behaviour of complex systems in several fields, such as Natural Language Processing (NLP), autonomous driving, and smart healthcare [5, 6]. ML is a branch of Artificial Intelligence (AI), which comprises the study of computer algorithms that are used to improve performance through automated recognition of patterns and regularities in data sets [5]. Technology giants such as Google, Microsoft, Facebook, Amazon, Nvidia, and others have invested heavily with their powerful computing resources to boost AI research, mainly aiming at breakthroughs and improving ML techniques. Such research efforts evince that ML models will improve both current and next generation of computing systems.

While emerging learning techniques fulfill a broad range of applications, the accuracy of underlying ML models comes at the expense of high computational and memory requirements for both the training and the inference phases. Training a ML model is space and computationally expensive due to the high set of parameters, which are iteratively refined over multiple executions. An inference model might still be complex due to the potentially high density of the input data (e.g., a high-resolution image) and the vast number of computations that need to be performed during its execution. Even with such computational requirements, recent works denote an emerging trend in creating inference versions of ML models that aim at enabling their execution in edge devices, which rely on some security, reliability, resource, and power constraints [7].

G. Abich et al.. *Early Soft Error Reliability Assessment of Convolutional Neural Networks Executing on Resource-Constrained IoT Edge Devices*, Synthesis Lectures on Engineering, Science. and Technology, https://doi.org/10.1007/978-3-031-18599-1_1

ML models are used in complex safety-critical systems (e.g., self-driving vehicles) and have, more recently, been incorporated in resource-constrained Internet of Things (IoT) edge devices. IoT edge devices combine embedded technologies such as wired and wireless communications, sensor and actuator devices, and the physical objects connected to the internet [8, 9]. Such systems are able to access raw data from different resources over the network and analyze a high set of information to extract knowledge [1]. Currently, the ML models most employed in resource-constrained IoT systems are the pre-trained Neural Networks (NN), such as Convolutional Neural Networks (CNN)s, since they do not perform the complex training phase on the target device [10]. For instance, CNNs are used to detect, identify, or classify patterns at the edge of IoT world [11]. However, apart from robust algorithms and high-performance hardware, executing such complex algorithms under resource-constrained devices is still a challenging task [12].

Aiming to enable the execution of such computationally intensive CNNs under resource-constrained IoT edge devices, researchers and industrial leaders are investigating software libraries and Application Programming Interfaces (API)s [13]. Such approaches rely on the development of lightweight and optimized NN kernels, allowing their execution on low-power and less efficient processors typically found in edge devices. Besides that, the inference of such kernels is based on reduced precision quantization schemes that exploits the trade-off between accuracy and data size to reduce the memory footprint [10]. In this sense, the main challenges to enable the execution of CNNs in such devices are the follows:

- (i) reduce the memory footprint, considering both application object code and CNN memory parameters;
- (ii) provide optimised and accurate NN inference models taking into account the architecture particularities, such as Single Instruction Multiple Data (SIMD);
- (iii) validate the resultant model in real world environment considering a target IoT edge device.

Although the broad applicability provided by such lightweight and optimized approaches, their changes need to meet performance and reliability requirements considering different environments.

Challenges in existing CNNs that targets the execution in IoT systems must consider several threats that affect both the reliability and the efficiency of the entire system [14]. The deployment of such models in market leader IoT edge devices requires robustness features, since soft errors and system malfunctions might have potentially fatal implications in safety-critical application areas [15]. For example, while the slight inaccuracy in ML models like NLP does not have any severe consequences, a small error in safety-critical applications like autonomous driving vehicles and smart healthcare can lead to catastrophic effects. In contrast with high-accuracy ML models, there is a significant need for robust CNN models that can generate reliable and trustworthy results in the presence of faults while also preserving safety and healthy. For this reason, software engineers must develop lightweight and reliable applications aiming to guarantee a fail-safe critical IoT edge system. In this sense, it is essential to evaluate the soft error susceptibility of such applications executing on resource-constrained IoT edge devices. With that in mind, researchers have started to investigate the impact of radiation-induced soft errors on the reliability of ML models.

In the reliability aspect, the soft error resilience is emerging as a key design metric due to the increasing susceptibility of electronic computer systems to the occurrence of soft errors caused by radiation effects [16]. In IoT edge computing, one open problem is how to reliably mine real-world data from a noisy and complex environment that confuses conventional CNNs [6]. The complexity of CNNs executing on IoT edge systems impose both software and hardware exploration challenges, including:

- (i) conduct a large number of Fault Injections (FI) campaigns within a reasonable time;
- (ii) provide engineers with detailed observation of a system's behavior in the presence of soft errors occurring both system execution (i.e., registers) and memory parameters (i.e., Flash and RAM);
- (iii) investigate the impact of soft errors in optimized kernel characteristics according to platform specific parameters;
- (iv) investigate the impact of soft errors in reduced precision NN memory parameters;
- (v) investigate the impact of soft errors when considering thread parallelism optimizations targeting existing NN kernels;
- (vi) identify relationships or associations between application characteristics and specific platform parameters in large data sets resulting from the fault campaigns.

The resulting scenario calls for faster and more efficient means to assess the soft error resilience of such complex systems with minimal overhead in time-to-market [17–22].

With that in mind, current researches are proposing virtual platform FI frameworks to cover such response time constraints [23–25]. Virtual Platforms (VP) frameworks have gained popularity over the years, not only in academia but also in many industrial sectors, due to their design flexibility, debugging, and simulation performance capacities. In this sense, works incorporated FI capability into VP frameworks [26, 27], enabling the analysis of complex software stacks and processor architectures at early design phases. However, to ensure the failsafe functionality of emerging resource-constrained IoT systems, engineers should be able to not only assess and identify, but also to promote efficient alternatives to mitigate the occurrence of soft errors [28]. Furthermore, while the gain in simulation speed is trivially observed in VP simulators based on Just-In-Time (JIT) dynamic binary translation, the soft error assessment consistency of underlying FI frameworks remains unclear. The preceding context motivates this Book, which aims at investigating the consistency of JIT-based FI techniques and use these tools to assess and mitigate the occurrence of soft errors in resource-constrained IoT devices executing CNN models.

This work first focuses on elucidating the consistency of the results gathered with the Soft error Fault Injection Analysis (SOFIA) framework [29], when compared to an Register-

Transfer Level (RTL) FI approach. With this consistency assessment, this book thus focuses on enhancing SOFIA capability by proposing two extensions: (*i*) a realistic fault classification according to CNN output predictions; and (*ii*) an automated FI technique that isolates specific memory sections in order to evaluate the impact of soft errors on different memory parameters of CNN models executing on resource-constrained IoT edge devices. The proposed extensions move SOFIA beyond the traditional frameworks, thereby furthering potential advantages that enable engineers to evaluate specific aspects of emerging CNN models. Furthermore, this work extends existing NN kernels employing parallel capabilities in order to evaluate the impact of thread parallelism on the performance and reliability of an CNN model executing on IoT edge device. Finally, this Book presents different case studies that evaluate the relative trade-off between different soft error mitigation techniques considering reliability, performance, and precision of the adopted CNN model.

1.1   Hypothesis to Be Demonstrated in This Book

This Book relies on three hypotheses:

- The first hypothesis of this Book is that virtual platform FI frameworks provide truthful results with significant performance gains when compared to the most classical ones. A fast and consistent FI tool enable engineers to ensure the accuracy of soft errors and failures to assess the reliability of complex computing systems (i.e., real software stacks, state-of-the-art Instruction Set Architectures (ISA)s), ensuring a realistic results at early design phases. Furthermore, the proof of this hypothesis allow us to enable us to provide a consistent assessment of the soft error reliability of CNN models, which would be unfeasible in traditional models due to the simulation time.
- With the use of architecture specific kernel optimizations as well as reduced precision to enable the execution of CNN models on IoT systems, engineers are more likely to identify meaningful relationships or associations between FI results, software optimizations, and parameters quantizations. In this regard, the second hypothesis of this book is to early investigate and identify the correlation between FI results, NN optimized kernels, and reduced precision parameters of CNNs executing on resource-constrained IoT edge devices. The proof of this hypothesis allow us to distinguish soft errors occurring in such models and apply lightweight and non-intrusive mitigation techniques to improve the reliability and lifetime of CNN models executing on resource-constrained IoT edge systems.
- When dealing with those parameters, the existing evaluation techniques in the literature are not sufficient to correlate the real impact of faults in the soft error reliability of CNN models considering the IoT device storage. With that in mind, the third hypothesis of this Book is that the reduced precision optimizations impacts not only in the CNN execution,

but also in the CNN code, parameters, and data stored in memory. The proof of such hypothesis allow us to isolate and identify the most vulnerable memory sections of a CNN deployed to a resource-constrained IoT edge device.

1.2 Book Goal

In order to address the reliability of ML models using VPs, the strategic goal of this Book is first to perform an in-depth and statistical significance soft error consistency evaluation of a JIT-based FI framework called SOFIA [20, 29]. This work aims to further contribute to the use of virtual platform FI frameworks in the design flow of safety-critical industry applications considering soft error reliability aspect. The second strategic goal of this Book is to combine existing features support by SOFIA (e.g., FI techniques) along with new tools and mitigation techniques aiming to assess the soft error reliability of CNN models executing on resource-constrained IoT edge devices considering different case studies, model optimizations (e.g., precision quantization, thread parallelism), processor models and ISAs. Finally, the third strategic goal of this book comprises to evaluate the impact of soft errors on memory units of IoT edge devices executing reduced precision ML models.

The following specific objectives should be fulfilled to accomplish the first strategic goal:

- Port several benchmarks from embedded and IoT domains, considering standard open-source Clang and GNU Compiler Collection (GCC) compilers and commercial Advanced RISC Machines (Arm) compilers, along with standard optimization flags to cover a wide range of probabilities in soft error consistency evaluation.
- Port FI techniques and scripts that enable us to trace, evaluate, and identify the particular source of errors in RTL processor descriptions.
- Evaluate the soft error assessment consistency of a VP FI tool w.r.t. a real single-core commercial processors at the RTL with a discrete event full system simulator.

The second goal requires the following task to be achieved:

- Port reduced and mixed precision CNN models for the SOFIA environment to enable the soft error assessment with a significant number of FI campaigns.
- Propose a realistic fault classification aiming to evaluate the real impact on the output parameters of CNN models.
- Investigate the impact of reduced precision and architecture specific optimizations on soft error reliability of CNN models executing on resource-constrained IoT systems.
- Employ the use of soft error software-based mitigation techniques to improve the reliability of CNN models executing on resource-constrained IoT systems.
- Assess both performance and reliability impact when using lightweight and non-intrusive software-based mitigation techniques in such CNN models.

- Evaluate the relative trade-off between soft error software-based mitigation techniques considering performance, reliability, and precision.

Finally, the third goal comprises the following milestones:

- Evaluate different aspects of CNN models executing on IoT systems using novel FI techniques that enable us to isolate specific system execution moments and memory sections.
- Investigate the impact of reduced and mixed precision memory parameters on the soft error reliability of the adopted CNN models.
- Investigate the impact of thread parallelism on soft error reliability of existing NN kernels.

1.3   Original Contributions of This Book

Figure 1.1 illustrates the main contributions of this work, which are joined into a soft error analysis flow that includes: the FI techniques and extensions, fault classification, and soft error mitigation techniques (fully described in Chap. 4); the cross compilation and platform simulators used to evaluate the consistency of the SOFIA framework (described in Chap. 5); the case studies and libraries adopted to asses soft error reliability of ML models executing on resource-constrained IoT edge devices (described in Sects. 6.1 and 6.2); the validation boards; the automated framework flow that allowed to evaluate different aspects of soft error reliability in a feasible time, i.e., Sects. 6.1.2 and 6.2.2 considering the ML model execution (e.g., isolated layers, target ISA optimization, precision bitwidth), Sects. 6.1.3 and 6.2.3 considering the ML model storage, Sect. 6.1.4 considering thread parallelism; and the adopted non intrusive soft error software-based mitigation techniques used as alternative to improve reliability in the adopted case studies with lower performance costs in Sect. 6.2.4.

The contributions of this Book are summarized as follows:

1.3.1   Evaluation of SOFIA Consistency w.r.t. RTL

The design flexibility, debugging and simulation performance capabilities of virtual platform frameworks led to the increase the popularity in both academia and industrial sectors. While the gain in simulation speed is trivially observed in VP simulators based on JIT dynamic binary translation, the soft error assessment consistency of underlying FI frameworks remains unclear. In order to investigate the soft error consistency of VP simulators, recent works demonstrate that higher level FI approaches provide more flexibility and simulation performance at the cost of accuracy, mostly resulting from the lack of microarchitecture and timing modelling aspects [19, 24, 30]. In this direction, one of the *main contributions* of this book is an in-depth and statistical significance soft error consistency evaluation of

SOFIA framework [29] against FI campaigns conducted with event-driven simulators (i.e., more realistic and accurate platforms) considering Arm Cortex-M processor architectures. Reference single-core FI campaigns are performed on RTL descriptions of Arm Cortex-M0 and M3 processors. Campaigns consider different open-source and commercial compilers as well as real software stacks including FreeRTOS kernel and 26 applications. Such contribution have already published at [22].

1.3.2   Early Soft Error Assessment of ML Models Executing on Resource-Constrained IoT Edge Devices

ML algorithms have provided straightforward solutions to a wide range of applications. The high computational demand of such algorithms limits their adoption in resource-constrained devices, typically relying on reduced memory footprint, low-power, and low performance processors. While performance improvement, customized, and reduced-precision implementations of ML models have been studied extensively, their susceptibility to soft errors caused by radiation particles is still an open research question. In this regard, the *second main contribution* of this book relies on the soft error reliability assessment of reduced and mixed precision CNNs executing on resource-constrained IoT edge devices. In order to cover a wide range of aspects from such models, this study evaluates the soft error reliability considering the fault impact in:

- isolated critical function layers;
- different processor architectures;
- reduced and mixed precision quantizations;
- isolated memory sections;
- multi-threaded execution.

In this contribution, we *propose* a more realistic fault classification to evaluate the impact of soft errors on the output probabilities of the target models. The evaluated results employs a FI technique that isolates the critical layers' functions of the adopted CNNs. The second case study [31] investigates the impact of precision bitwidth on the soft error reliability of the MobileNet [32] CNN developed based on the CMSIS-based Mixed precision API for Neural Networks (CMix-NN) library [33] and executed on an Arm Cortex-M processor. In the subsequent study, we *propose* an extension to the physical memory FI technique available in SOFIA [29] that isolates the memory sections, evaluating and comparing the fault impact on the object code and the CNN weights, bias, and buffers. Furthermore, our late approach aims to assess the soft error reliability of a multi-threaded version of a CNN model developed based on the Arm CMSIS API for Neural Networks (CMSIS-NN) kernels considering 1, 2 and 4 cores. Such contributions have already published at [31, 34–37].

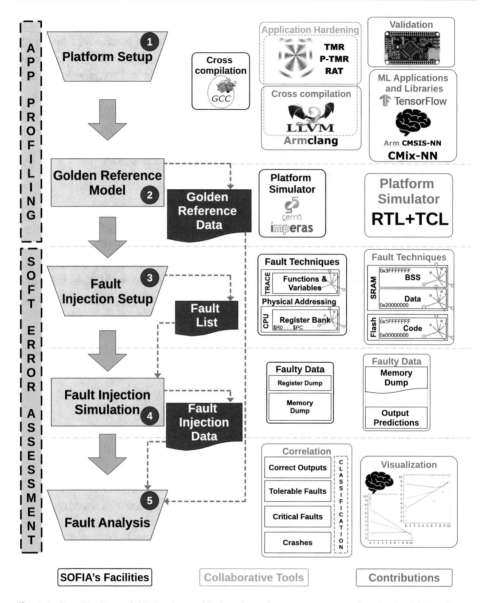

Fig. 1.1 Contributions of this book considering the soft error assessment flow in the SOFIA framework

1.3.2.1 Applying Lightweight Soft Error Mitigation Techniques to Embedded Mixed Precision Deep Neural Networks

While CNNs' precision and performance can vary and are essential, it is also vital to deploy trained models that provide high reliability at low cost. To achieve an unyielding reliability and safety level, it is imperative to provide electronic computing systems with appropriate mechanisms to tackle soft errors. In this sense, the *third main contribution* of this Book is to investigate the relationship between soft errors, performance, and model accuracy while promoting the use of a Register Allocation Technique (RAT) [28] that allocates the critical CNN function/layer to a pool of specific general-purpose processor registers. This study comprises an extensive soft error assessment of the MobileNet model considering 2, 4, and 8 precision bitwidth variations running on an Arm Cortex-M processor. Such results have already published [35].

1.3.2.2 Consistent and Extensive Evaluation

In order to provide a consistent and extensive evaluation, the results comprise more than 14.8 millions of FIs considering distinct case studies, architectures, number of cores, OS, parallelization libraries. Different from other works, the promoted framework flow uses a realistic software stack comprising bare-metal applications and running over an operating system (e.g., Linux and FreeRTOS), parallelization libraries (e.g., OpenMP), and ML libraries (e.g., CMSIS-NN and CMix-NN). Furthermore, to reinforce the experiments, all adopted case study applications (i.e., benchmarks and CNNs) have been validated by running them on a real board or on the available RTL description, which also provide some performance measurements.

1.3.3   Legacy of Tools Integrated into the SOFIA Framework

In addition to the evaluation of soft error consistency the contributions of this work are fully integrated into SOFIA automated soft error analysis flow, which leverages the high simulation performance of M*DEV [38] to efficiently acquire representative error/failure-related data considering state-of-the-art single and multi-core processor architectures, from ARMv6-M to recent ARMv8-A. Such contributions also comprise the support to:

- compiler scripts considering GCC, Clang, and Arm compilers;
- ML libraries and models targeting resource-constrained IoT edge devices;
- TCL scripts to assess the soft error reliability in RTL processor descriptions;
- a fault injection technique that isolates memory sections;
- a fault classification considering the critical faults affecting the output probabilities of ML models.

Furthermore, the visualization tools have been improved to meet the proposed assessments to correlating faulty results and find some particular behavior according to each case study.

1.4   Book Outline

This Book is organized into seven Chapters:

In this chapter—Introduction: this Chapter introduces the current issues found in the literature to enable a consistent reliability assessment of ML models executing on resource-constrained IoT edge devices and summarizes the contributions of this Book. The following paragraphs present a succinct Book summary Chapter by Chapter.

Chapter 2—Background: this Chapter presents the required background in works related to ML models (Sect. 2.1) and Radiation induced soft errors (Sect. 2.2).

Chapter 3—Related Works: this Chapter presents the works related to the challenges of evaluating soft errors in ML models using VPs. In this sense, Sect. 3.1 presents the state-of-the-art on soft error assessment using VPs. Next, Sect. sec:stateml presents the state-of-the-art on soft error assessment of ML models considering different FI approaches. Finally, we place the contributions of this work regarding the ones found in the literature in Sects. 3.2.2 and 3.1.1.

Chapter 4—Methodology: this Chapter describes the adopted methodology in this study considering the FI approaches, the fault classification, and the assessment metrics. First, Sect. 4.1 details the adopted FI approaches and their different level of accuracy. Then, Sects. 4.1.1, 4.1.2.1, and 4.1.2.2 details each adopted FI modules for RTL, gem5, and Open Virtual Platform Simulator (OVPsim) respectively. Section 4.2 details the proposed fault classification according the ones existing in the literature. Finally, Sect. 4.3 presents the assessment metrics used to asses the soft error reliability from the adopted case studies, and Sect. 4.3.1 present a brief description of the mitigation techniques promoted in this work.

Chapter 5—Results on Soft Error Consistency Assessment: this Section presents the conducted experiments considering different FI approaches to evaluate the consistency of SOFIA against RTL and gem5 FI approaches. Section 5.1 presents the results of the soft error consistency assessment for single-core processors. In each of those analyses, we present a detailed consistency evaluation addressing different aspects of adopted architectures and simulators.

Chapter 6—Soft Error Reliability Assessment of ML Models on Resource-Constrained IoT Edge Devices: this Chapter presents the proposed case studies considering the execution of ML models in resource-constrained IoT edge devices. Sections 6.1 and 6.2 details the ML kernels and APIs used to deploy the adopted case studies. Then, following sub sections comprise the results from conducted FI campaigns considering different soft error reliability aspects on the execution of CNN models in resource-constrained IoT edge devices.

Chapter 7—Conclusions: this Chapter summarizes the contributions of this while high-lights the results published in significant journal papers and peer-reviewed international conferences. Finally, this Chapter also points out the possible future works enabled by the contributions reported in this Book.

References

1. Mahdavinejad, M.S., Rezvan, M., Barekatain, M., Adibi, P., Barnaghi, P., Sheth, A.P.: Machine learning for internet of things data analysis: a survey. Digit. Commun. Netw. **4**(3), 161–175 (2018). https://doi.org/10.1016/j.dcan.2017.10.002
2. CISCO: Cisco annual internet report white paper (2021). https://www.cisco.com/c/en/us/solutions/executive-perspectives/annual-internet-report
3. Labrinidis, A., Jagadish, H.V.: Challenges and opportunities with big data. Proc. VLDB Endow. **5**(12), 2032–2033 (2012). https://doi.org/10.14778/2367502.2367572
4. Najafabadi, M.M., Villanustre, F., Khoshgoftaar, T.M., Seliya, N., Wald, R., Muharemagic, E.: Deep learning applications and challenges in big data analytics. J. Big Data **2**(1), 1 (2015). https://doi.org/10.1186/s40537-014-0007-7
5. Samuel, A.L.: Some studies in machine learning using the game of checkers. IBM J. Res. Dev. **3**(3), 210–229 (1959). https://doi.org/10.1147/rd.33.0210
6. Li, H., Ota, K., Dong, M.: Learning IoT in edge: deep learning for the internet of things with edge computing. IEEE Netw. **32**(1), 96–101 (2018). https://doi.org/10.1109/MNET.2018.1700202
7. Chen, J., Ran, X.: Deep learning with edge computing: a review. Proc. IEEE **107**(8), 1655–1674 (2019). https://doi.org/10.1109/JPROC.2019.2921977
8. Atzori, L., Iera, A., Morabito, G.: The internet of things: a survey. Comput. Netw. **54**(15), 2787–2805 (2010). https://doi.org/10.1016/j.comnet.2010.05.010
9. Cecchinel, C., Jimenez, M., Mosser, S., Riveill, M.: An architecture to support the collection of big data in the internet of things. In: 2014 IEEE World Congress on Services, pp. 442–449 (2014). https://doi.org/10.1109/SERVICES.2014.83
10. Jacob, B., Kligys, S., Chen, B., Zhu, M., Tang, M., Howard, A., Adam, H., Kalenichenko, D.: Quantization and training of neural networks for efficient integer-arithmetic-only inference. In: Proceedings of the IEEE Conference on Computer Vision and Pattern Recognition (CVPR), pp. 2704–2713 (2018). https://doi.org/10.1109/CVPR.2018.00286
11. Khan, S., Muhammad, K., Mumtaz, S., Baik, S.W., de Albuquerque, V.H.C.: Energy-efficient deep CNN for smoke detection in foggy IoT environment. IEEE Internet Things J. **6**(6), 9237–9245 (2019). https://doi.org/10.1109/JIOT.2019.2896120
12. Adi, E., Anwar, A., Baig, Z., Zeadally, S.: Machine learning and data analytics for the IoT. Neural Comput. Appl. **32**(20), 16205–16233 (2020). https://doi.org/10.1007/s00521-020-04874-y
13. Lee, J., Stanley, M., Spanias, A., Tepedelenlioglu, C.: Integrating machine learning in embedded sensor systems for internet-of-things applications. In: 2016 IEEE International Symposium on Signal Processing and Information Technology (ISSPIT), pp. 290–294 (2016). https://doi.org/10.1109/ISSPIT.2016.7886051
14. Punithavathi, P., Geetha, S., Karuppiah, M., Islam, S.H., Hassan, M.M., Choo, K.K.R.: A lightweight machine learning-based authentication framework for smart IoT devices. Inf. Sci. **484**, 255–268 (2019). https://doi.org/10.1016/j.ins.2019.01.073
15. Kriebel, F., Rehman, S., Hanif, M.A., Khalid, F., Shafique, M.: Robustness for smart cyber physical systems and internet-of-things: from adaptive robustness methods to reliability and

security for machine learning. In: 2018 IEEE Computer Society Annual Symposium on VLSI (ISVLSI), pp. 581–586 (2018). https://doi.org/10.1109/ISVLSI.2018.00111

16. Baumann, R.: Soft errors in advanced computer systems. IEEE Des. Test Comput. **22**(3), 258–266 (2005). https://doi.org/10.1109/MDT.2005.69

17. Mukherjee, S., Emer, J., Reinhardt, S.: The soft error problem: an architectural perspective. In: 11th International Symposium on High-Performance Computer Architecture, pp. 243–247 (2005). https://doi.org/10.1109/HPCA.2005.37

18. Li, X., Shen, K., Huang, M.C., Chu, L.: A memory soft error measurement on production systems. In: USENIX Annual Technical Conference, pp. 275–280 (2007). https://dl.acm.org/doi/10.5555/1364385.1364406

19. Chatzidimitriou, A., Bodmann, P., Papadimitriou, G., Gizopoulos, D., Rech, P.: Demystifying soft error assessment strategies on ARM CPUs: microarchitectural fault injection versus neutron beam experiments. In: International Conference on Dependable Systems and Networks (DSN), pp. 26–38 (2019). https://doi.org/10.1109/DSN.2019.00018

20. da Rosa, F.R., Garibotti, R., Ost, L., Reis, R.: Using machine learning techniques to evaluate multicore soft error reliability. IEEE Trans. Circuits Syst.-I: Regul. Pap. **66**(6), 2151–2164 (2019). https://doi.org/10.1109/TCSI.2019.2906155

21. Bodmann, P., Papadimitriou, G., Rech Junior, R.L., Gizopoulos, D., Rech, P.: Soft error effects on arm microprocessors: early estimations versus chip measurements. IEEE Trans. Comput. 1 (2021). https://doi.org/10.1109/TC.2021.3128501

22. Abich, G., Garibotti, R., Bandeira, V., da Rosa, F., Gava, J., Bortolon, F., Medeiros, G., Moraes, F.G., Reis, R., Ost, L.: Evaluation of the soft error assessment consistency of a JIT-based virtual platform simulator. IET Comput. Digit. Tech. **15**(2), 125–142 (2021). https://doi.org/10.1049/cdt2.12017

23. Hari, S.K.S., Adve, S.V., Naeimi, H., Ramachandran, P.: Relyzer: Exploiting application-level fault equivalence to analyze application resiliency to transient faults. In: Proceedings of the Seventeenth International Conference on Architectural Support for Programming Languages and Operating Systems, pp. 123–134. Association for Computing Machinery (2012). https://doi.org/10.1145/2150976.2150990

24. Kaliorakis, M., Tselonis, S., Chatzidimitriou, A., Foutris, N., Gizopoulos, D.: Differential fault injection on microarchitectural simulators. In: International Symposium on Workload Characterization (IISWC), pp. 172–182 (2015). https://doi.org/10.1109/IISWC.2015.28

25. Tanikella, K., Koy, Y., Jeyapaul, R., Lee, K., Shrivastava, A.: gemV: a validated toolset for the early exploration of system reliability. In: 2016 IEEE 27th International Conference on Application-Specific Systems, Architectures and Processors (ASAP), pp. 159 – 163 (2016). https://doi.org/10.1109/ASAP.2016.7760786

26. Parasyris, K., Tziantzoulis, G., Antonopoulos, C.D., Bellas, N.: GemFI: A fault injection tool for studying the behavior of applications on unreliable substrates. In: International Conference on Dependable Systems and Networks (DSN), pp. 622–629 (2014). https://doi.org/10.1109/DSN.2014.96

27. da Rosa, F., Bandeira, V., Reis, R., Ost, L.: Extensive evaluation of programming models and ISAs impact on multicore soft error reliability. In: Design Automation Conference (DAC), pp. 1–6 (2018). https://doi.org/10.1145/3195970.3196050

28. Gava, J., Reis, R., Ost, L.: RAT: A lightweight system-level soft error mitigation technique. In: IFIP/IEEE International Conference on Very Large Scale Integration (VLSI-SOC), pp. 165–170 (2020). https://doi.org/10.1109/VLSI-SOC46417.2020.9344080

29. Bandeira, V., Rosa, F., Reis, R., Ost, L.: Non-intrusive fault injection techniques for efficient soft error vulnerability analysis. In: 2019 IFIP/IEEE 27th International Conference on Very Large

Scale Integration (VLSI-SoC), pp. 123–128 (2019). https://doi.org/10.1109/VLSI-SoC.2019.892037

30. Rosa, F., Ost, L., Reis, R., Davidmann, S., Lapides, L.: Evaluation of multicore systems soft error reliability using virtual platforms. In: International New Circuits and Systems Conference (NEWCAS), pp. 85–88 (2017). https://doi.org/10.1109/NEWCAS.2017.8010111

31. Abich, G., Reis, R., Ost, L.: The impact of precision bitwidth on the soft error reliability of the mobilenet network. In: IEEE Latin American Symposium on Circuits and Systems (LASCAS), pp. 1–4 (2020). https://doi.org/10.1109/LASCAS51355.2021.9667153

32. Howard, A.G., Zhu, M., Chen, B., Kalenichenko, D., Wang, W., Weyand, T., Andreetto, M., Adam, H.: MobileNets: efficient convolutional neural networks for mobile vision applications (2017). arXiv:1704.04861. https://doi.org/10.48550/arXiv.1704.04861

33. Capotondi, A., Rusci, M., Fariselli, M., Benini, L.: CMix-NN: Mixed low-precision CNN library for memory-constrained edge devices. IEEE Trans. Circuits Syst. II: Express Briefs **67**(5), 871–875 (2020). https://doi.org/10.1109/TCSII.2020.2983648

34. Abich, G., Gava, J., Reis, R., Ost, L.: Soft error reliability assessment of neural networks on resource-constrained IoT devices. In: IEEE International Conference on Electronics, Circuits and Systems (ICECS), pp. 1–4 (2020). https://doi.org/10.1109/ICECS49266.2020.9294951

35. Abich, G., Gava, J., Garibotti, R., Reis, R., Ost, L.: Applying lightweight soft error mitigation techniques to embedded mixed precision deep neural networks. IEEE Trans. Circuits Syst. I: Regul. Pap. **68**(11), 4772–4782 (2021). https://doi.org/10.1109/TCSI.2021.3097981

36. Abich, G., Garibotti, R., Reis, R., Ost, L.: The impact of soft errors in memory units of edge devices executing convolutional neural networks. IEEE Trans. Circuits Syst. II: Express Briefs **69**(3), 679–683 (2022). https://doi.org/10.1109/TCSII.2022.3141243

37. Abich, G., Gava, J., Garibotti, R., Reis, R., Ost, L.: Impact of thread parallelism on the soft error reliability of convolution neural networks. In: IEEE Latin American Symposium on Circuits and Systems (LASCAS), pp. 1–4 (2022). https://doi.org/10.1109/LASCAS53948.2022.9789088

38. Imperas: DEV—Virtual Platform Development and Simulation (2021). https://www.imperas.com/dev-virtual-platform-development-and-simulation/

Background in ML Models and Radiation Effects   2

This Chapter aims to introduce some necessary backgrounds and works related to this Book, the soft error assessment of ML models applied to IoT systems. First, Sect. 2.1 presents a background in ML models and the challenges to enable the execution of such models in IoT edge devices. Further, Sect. 2.2 addresses the basic concepts regarding radiation-induced errors and their impact on electronic computing system devices.

2.1   Basic Concepts of ML Models

ML is a special branch of artificial intelligence that acquires knowledge from training data based on known facts without explicit programming [1]. As described by [2], ML is a *"Field of study that gives computers the ability to learn without being explicitly programmed"*. Learning is a matter of finding statistical regularities or other patterns in the data, and the performance analysis of such learning models can give insight into relative difficulty of learning in different environments [3]. ML models are organized into a taxonomy presented in Fig. 2.1 [4]. The literature classifies ML models into three broad categories such as: supervised learning, unsupervised learning, and reinforcement learning.

Supervised learning is the learning method where the instances are labeled in the training phase generating a function that maps inputs to desired outputs. *Unsupervised learning* is a type of ML model, which models a set of inputs without labeled examples. *Reinforcement learning* means a computer interacting with an environment to achieve a certain goal considering a policy of how to act given an observation of the real world scenario. Every action has some impact in the environment, and the environment provides feedback that guides the learning algorithm.

According to reviewed works [1, 5–7], supervised and unsupervised learning have been and are still widely applied in smart data analysis. In this sense, the following subsections

© The Author(s), under exclusive license to Springer Nature Switzerland AG 2023   15
G. Abich et al., *Early Soft Error Reliability Assessment of Convolutional Neural Networks Executing on Resource-Constrained IoT Edge Devices*, Synthesis Lectures on Engineering, Science, and Technology, https://doi.org/10.1007/978-3-031-18599-1_2

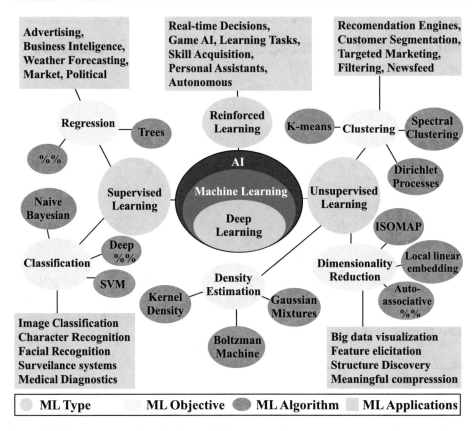

Fig. 2.1 Different types and applications of ML exploited for solving computer science problems. Adapted from [4]

present a review focused on these ML types (Sects. 2.1.2 and 2.1.1) thus focusing on ML models commonly used in IoT devices (Sect. 2.1.3).

2.1.1 Unsupervised Learning

Unsupervised learning techniques are regarded as self-learning algorithms that possess the capacity to explore and locate the previously unknown patterns in a dataset [8]. While supervised learning (Sect. 2.1.2) requires a labeled dataset, in unsupervised learning algorithms the training data consists of a set of input vectors without any corresponding target values [9]. Such learning algorithms are left to their own devises using a set of statistical tools to discover and represent interesting patterns from data structures [10].

Some of the most common unsupervised learners are: Cluster analysis (K-means clustering, Fuzzy clustering), Hierarchical clustering, Self-organizing map, Apriori algorithm,

Eclat algorithm, and Outlier detection. Unsupervised learning algorithms can be further grouped into the following three categories: clustering, dimensionality reduction, and density estimation (e.g., Fig. 2.1).

Clustering is the most common unsupervised learning problem, which categorizes the observed dataset into groups that maximize some similarity criterion, equivalently dissimilar to the objects belonging to other clusters. Clustering algorithms may be classified as Exclusive (Fig. 2.2a), Hierarchical (Fig. 2.2b), Overlapping (Fig. 2.3a), and Probabilistic Clustering (Fig. 2.3b). In exclusive or partitioning clustering, data are grouped in such a way that one data can belong to one cluster only [11]. Hierarchical clustering technique considers every data as a cluster and the iterative unions between the two nearest clusters reduce the number of clusters [12]. There are two types of hierarchical clustering, Divisive and Agglomerative. In the divisive (top-down) clustering method, the algorithm assigns all of the observations to a single cluster and then partition the cluster to two least similar clusters. Thus, the algorithm proceed recursively on each cluster until there is one cluster for each observation. In turn, agglomerative or bottom-up clustering method performs the opposite.

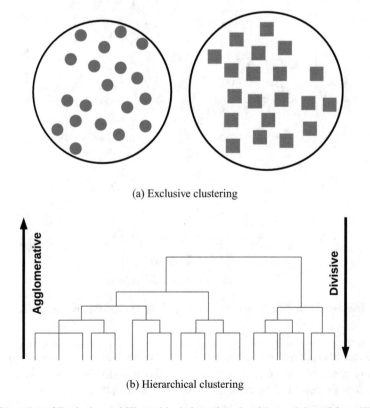

(a) Exclusive clustering

(b) Hierarchical clustering

Fig. 2.2 Examples of Exclusive and Hierarchical clustering algorithms. Adapted from [9]

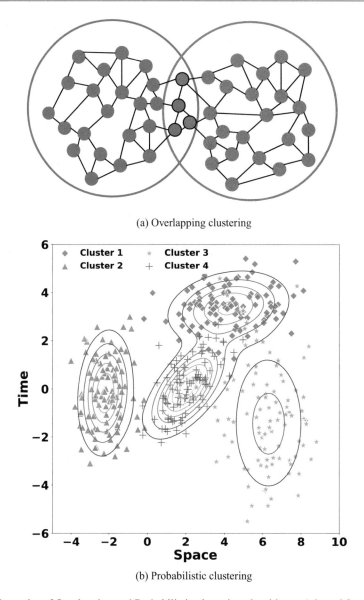

(a) Overlapping clustering

(b) Probabilistic clustering

Fig. 2.3 Examples of Overlapping and Probabilistic clustering algorithms. Adapted from [9]

The overlapping technique use fuzzy sets to cluster data where each point may belong to two or more clusters with separate degrees of membership [13, 14]. Probabilistic clustering technique uses probability distribution to create the clusters [15, 16].

Dimensionality reduction transforms a high-dimensional set of variables into a low-dimensional space where the resultant representation retains some meaningful properties of

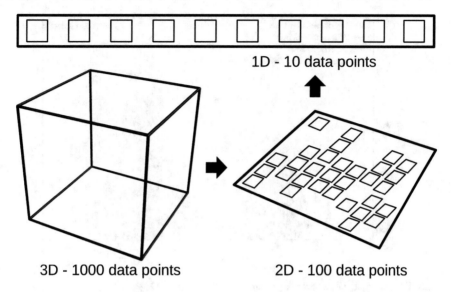

1D - 10 data points

3D - 1000 data points   **2D - 100 data points**

Fig. 2.4 Example of dimensionality reduction using matrix decomposition technique. Adapted from [22]

the original data, ideally close to its intrinsic dimension [17]. For instance, Fig. 2.4 shows an example of matrix decomposition data correlation used to reduce a 3D matrix to a single data vector. Such learning technique correlates the number of features present in the dataset in order to remove redundant information, reducing model's complexity, and avoid overfitting [18]. In this sense, dimensionality reduction splits into two main categories: *feature selection* and *feature extraction* [19]. In feature selection, the algorithm select a subset of features of the original dataset, aiming to obtain a model capable of automatically selecting the subset of features most relevant to a faced problem [20]. Feature extraction derives the information from the original dataset to reduce the number of features and build a new feature subspace [21]. In contrast to the feature selection, the output features will not be the same as the originals [19]. In this sense, the extracted features are combinations of the original features, compressed in a way that they will retain the most relevant information.

Density estimation uses statistical models to find an underlying probability distribution that gives rise to the observed variables [23]. For example, Fig. 2.5 shows the kernel density estimation for three choices of kernels considering as input 100 points from a bimodal distribution. The goal in such learning problems is to model the underlying structure or distribution in the data in order to identify the inherent structures of a dataset without using explicitly-provided labels [24]. Some of the most popular and useful density estimation techniques are mixture models such as Gaussian mixtures, and neighbor-based approaches (e.g., kernel density estimation) [25]. Gaussian mixtures are discussed more fully in the context of clustering, because the technique is also useful as an unsupervised clustering scheme.

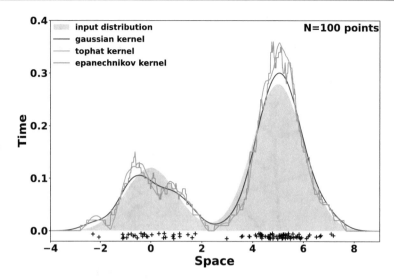

Fig. 2.5 1D Kernel density estimation example. Adapted from [26]

2.1.2 Supervised Learning

Supervised learning is the most common subbranch of ML [3]. The objective of such learning method is to learn how to predict the appropriate output vector for a given input vector. The supervised learning means that the training phase needs supervisor interaction. During the training phase, the algorithm will search for patterns in the data that correlate with the desired outputs. The training data will consist of inputs paired with the correct outputs. After training, a supervised learning algorithm will take in new unseen inputs and will determine which label the new inputs will be classified as based on prior training data.

There are several supervised learning techniques known in the literature such as: Artificial Neural Network (ANN), Bayesian Statistics, Gaussian Process Regression, Lazy learning, Nearest Neighbor, Support Vector Machine (SVM), Hidden Markov Model, Bayesian Networks, Decision Trees (C4.5, ID3, CART, Random Forrest), K-Nearest Neighbor (kNN), Boosting, Ensembles classifiers (Bagging, Boosting), Linear Classifiers (Logistic regression, Fisher Linear discriminant, Naive Bayes classifier, Perceptron), and Quadratic classifiers.

According to [28], supervised learning can be split into two subcategories: *classification* and *regression*. Applications where the target labels consist of a finite number of discrete categories are known as classification tasks [29]. Cases where the target labels are composed of one or more continuous variables are known as regression tasks [30]. In both regression and classification, the goal is to find specific relationships or structure in the input data that allow us to effectively produce correct output data. In this sense, a correct output is determined entirely from the training data. However, noisy or incorrect data labels can reduce the effectiveness of a given model in real-world situations. Thus, to provide efficient data labels, the main considerations when conducting supervised learning are the model complexity [29, 31] and the bias-variance trade-off [32–34].

In ML, the model complexity often refers to the number of features or terms included in a given predictive model. High complexity models are harder to interpret and have a high probability of overfitting, consequently these models are computationally expensive. Overfitting refers to learning a function that overly fits the training data, and essentially captures the noise and outliers. Underfitting happens when a model is unable to capture the underlying pattern of the data. As consequence, the model will not accurately predict cases that are not represented by the training data. There are some methods to control or reduce model complexity including linear modeling and subset selection, regularization, and dimensionality reduction. Such methods, essentially, keep all features, but reduce (or penalize) the effect of some features on the model's predicted values.

Figure 2.6 shows the bias-variance trade-off, where the center of the target is a model that perfectly predicts correct values. Bias is the difference between the average model's prediction and the correct value which must be predicted. Variance is the variability of the model's prediction for a given data point or a value that shows the dispersion of the dataset. While models with high bias oversimplify the trained model, models with high variance increases the model complexity. In both cases the resultant model leads to high error rates on the test data. This trade-off must find the right/good balance without overfitting and underfitting the data.

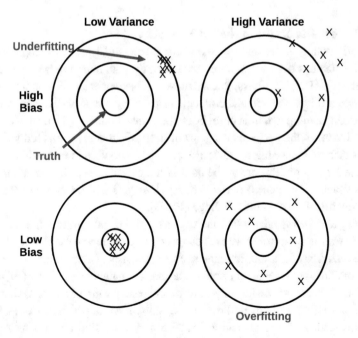

Fig. 2.6 Bias and variance using bulls-eye diagram. Adapted from [27]

Fig. 2.7 Model complexity
curve defined by Eq. (2.1) in
[27]

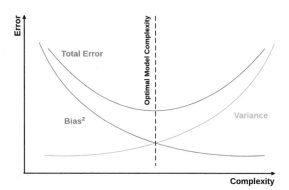

$$Total\ Errors = Bias^2 + Variance + Irreducible\ Error \qquad (2.1)$$

An optimal balance of bias and variance would never overfit or underfit the trained model. To build a good model, it is necessary to find a good balance between the bias and the variance such that it minimizes the total error defined by Eq. (2.1). The Total Errors is the sum of $Bias^2$, variance and the irreducible error. Irreducible error is the error that cannot be reduced by creating good models (i.e., amount of noise in training data). Finally, Fig. 2.7 shows the bias-variance trade-off curve where the center represents the optimal model complexity.

2.1.2.1 Pattern Recognition Neural Networks

Supervised Learning is the main focus of current practical pattern recognition applications [35], which has been more widespread recently with neural networks such as Deep Neural Networks (DNNs), ANNs, Recurrent Neural Networks (RNN)s, and CNNs. Neural Networks are artificial mathematical models inspired by a neuron-like structures used to process the data, aiming to mimic or replicate the human brain's sophisticated functionality. Figure 2.8 illustrates the performance gains of deep ML models compared with the traditional ones that are increasing the underlying models' adoption in real-world applications. Such different types of NNs are at the core of the current Deep ML solutions, powering applications such as Unmanned Aerial Vehicles (UAV)s [36, 37], self-driving cars [38, 39], speech recognition [40, 41], and IoT Edge [42, 43].

According to [35], a standard NN consists of a set of elementary and connected processing units, also known as neurons, where each one produces a sequence of real-valued activations. In this sense, an ANN (or a simple neural network) consists of a sequence of interconnected neuron layers, an input layer, a set of hidden layers, and a final output layer [44, 45]. Such a combination of simple processing units (e.g., nodes, perceptrons, or neurons) aims to acquire knowledge from the environment through a learning process, which stores the knowledge from the connection of neurons and their interactions [46]. Furthermore, an ANN is also

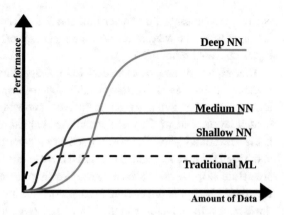

Fig. 2.8 The performance of Deep ML approaches compared with traditional ML

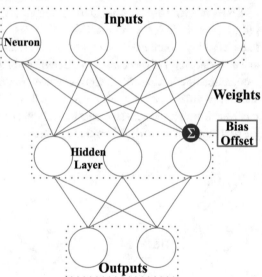

Fig. 2.9 Example of an ANN Architecture

known as a group of multiple perceptrons/neurons at each layer or as a Feed-Forward NN since the inputs are processed sequentially in the forward direction.

Figure 2.9 shows a typical ANN architecture, where lines connecting neurons are also shown. Each connection is defined as the weights, which are the features extracted at the training phase. Moreover, the bias is an additional input into the next layer to guarantee that even when all the inputs are zeros there will still be an activation in the next neuron. In this manner, [46] mathematically define a NN by the Eq. (2.2), which denotes the output, hi, of a neuron i in the hidden layer

$$hi = \sigma \left(\sum_{N}^{j=1} V_{ij} x_j + T_i^{hid} \right) \tag{2.2}$$

where $\sigma()$ represents the transfer function (i.e., activation), N is the number of neurons at the input layer, the weights are V_{ij}, the input neurons receive x_j as inputs, and the hidden layers threshold terms are T_i^{hid}.

One of the advantages of an ANN is its capability of learning any nonlinear function, popularly known as Universal Function Approximation. These types of NNs are able to learn weights that map any input to the output considering the learning process through the hidden layers. One of the main advantages behind universal approximation in the ANNs is the activation layers. The transfer/activation layer goal is to both introduce non-linearity and bound the neuron values to avoid NN freeze due to divergent neurons. Although ANNs are nature adaptive and have enhanced learning capacity, it is hard to train these NNs since they need a large amount of data. Furthermore, the number of trainable parameters increases drastically with an increase in the inputs' size (e.g., image or dataset).

RNNs inherit the same principles as ANNs and CNNs, about learning from training data [46]. However, different from ANNs and CNNs, RNNs use similar architecture to acquire knowledge from both past and input data, overcoming the weaknesses of the feed-forward, and artificial NNs [47]. For instance, Google and Apple use RNNs in practical voice/speech recognition applications (e.g., Google voice search and Apple's Siri) [48, 49]. Figure 2.10 shows the difference between an ANN and a RNN, which is related to the time aware recurrent learning [50]. Instead of using only the training data, it uses the recurrent data from the past inputs to make decisions during its execution. Nevertheless, RNNs have an in-depth

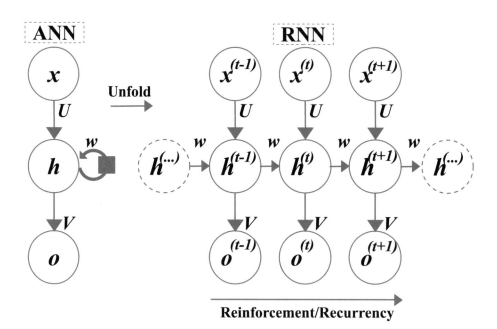

Fig. 2.10 Example of an RNN Architecture. Adopted from [50]

understanding of sequences and their context, contrasting with other NNs. Consequently, this type of neural network can also fit as reinforcement learning. However, these features turn this type of NN slow and complex to train. Moreover, the interdependence between layers makes it unfeasible to stack or parallelize RNNs.

The main role of CNNs in current computing systems is to act in video and image recognition [51–53], recommendation systems [54–57], and image analysis and classification [58–64]. In this sense, one can assume that, essentially, computer vision CNNs scale the process by using linear algebra principles to identify meaningful information from visual inputs, images, and videos. CNNs are based on three main layers: Convolutional layer, Pooling layer, and Fully-connected layer [65]. The convolutional layer preserves the spatial relationship between pixels by learning image features using small squares of input data. Next, the Pooling layer reduces processing dimensions between layers, thus increasing the performance along with the CNN execution. Additional to those layers, CNNs use activation and flattening normalization functions at the end of convolutional and pooling layers, respectively. Thus, at the end of the CNN, the Fully-connected layer performs the image classification, where each pixel or number/point is considered a separate neuron, just like a NN.

Figure 2.11 shows an example of the CNN computation that is performed through the layers. In this NN model, the network is able to recognize complex features, such as object shapes, as it processes images/frames through layers. Consequently, the learning complexity in understanding the image increases through the layers, which means the first layers are used to interpret simple features in the input image (e.g., as its edges and colors) while deeper layers are able to identify the target object. Although the challenges of using CNNs involve dealing with data variance during very complex training, the resulting dataset allows CNN to detect features automatically without human assistance. Consequently, these characteristics encourage the adoption of these NN models in IoT edge computing. However, such an approach brings new challenges regarding complexity (e.g., number of network layers), dataset size and low resources availability found in edge devices.

2.1.3 ML for the IoT Edge Devices

The size of computing devices in current technology nodes is largely reducing while their computing capability is drastically improving. Such resultant increase in computing power and advancements have made it possible to map and process much larger and deeper neural networks than was possible in earlier generations [67]. These upgrades encouraged the current shift of ML models to IoT environment, aiming to explore the execution of smart applications on resource-constrained devices [68]. In contrast to other ML models shown in Fig. 2.1 that must be highly tuned to solve specific problems and need a plenty of rules for successful operation, IoT edge deep inference techniques must deal with the constraints of power, performance, and resources inherent to such devices [4].

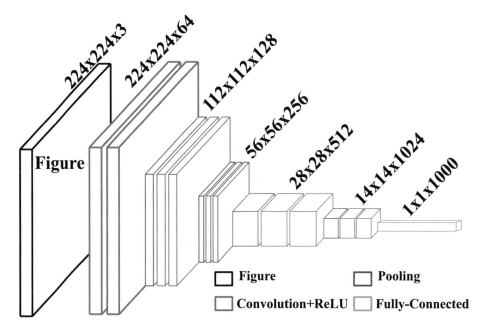

Fig. 2.11 Example of a CNN computation layers. Adopted from [66]

Figure 2.12 shows the trade-off between the techniques used to promote deep inference in HPC and resource-constrained devices. While the HPC environment capabilities support hardware acceleration, multi-core execution, and optimized instructions, to enable deep inference tasks on IoT edge side, there are two main considerations: architecture specific optimized software libraries and model pruning. One of the important aspects in the industrial IoT is the response time of such systems, which means that the IoT systems using ML models tend to be moved to the edge of the network. Currently, researchers aim to achieve a feasible performance while using optimizations allowed by the target ISAs. In this sense, to provide the execution of Deep Neural Network (DNN) models on the underlying devices, software libraries and APIs have been proposed [70–74]. Such libraries/APIs are devoted to streamlining the design and development of embedded deep learning-based applications through the fine-tuning of pre-trained network models, thus enabling their efficient execution in edge-computing platforms [1, 75]. In additional, to add more performance to IoT platform researchers also tried to make the ML models lighter with a similar level of performance and accuracy. Such approaches aim to reduce the number of inference parameters (e.g., weights and bias) to reduce the memory requirements from deep inference models while achieving such a performance improvements [76, 77].

The resultant trade-off between performance and resource constraints shifts the deep inference models to fixed point integer quantization methods. Such techniques aim to reduce the size of the parameters from deep neural networks and improve inference latency and

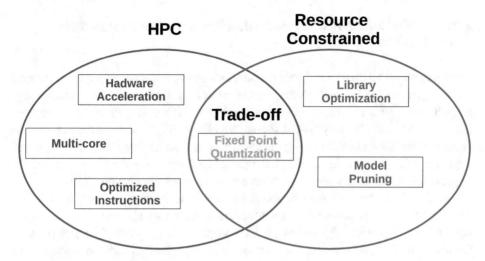

Fig. 2.12 Trade-off between approaches used to enable deep inference in resource-constrained IoT Edge devices. Adapted from [69]

throughput by taking advantage of high throughput integer instructions [78]. In this sense, researchers already found that quantized models reach almost the same level precision when considering reduced precision of NN parameters, mostly the weights [79, 80]. While the resultant precision is degraded (e.g., < 1% in MobileNet), the memory overhead is drastically reduced by 75% when using integer 8-bit parameters instead of 32-bit respectively. Moreover, with the cooperation of hardware, the quantization can also lead to faster model execution speed. Considering IoT applications targeting Arm processor, a category of SIMD instructions is introduced to accelerate the add and multiply operations [71, 74], which are the essential elements in ML model execution.

In the domain of resource-constrained devices one can find many implementations of ML models on mobile and embedded devices that cooperate with the cloud computing [67]. For the time being, the majority of embedded trained models and their inference engines have been evaluated only according to their accuracy and performance over a given dataset [69]. In contrast, as DNN become increasingly common in mission-critical applications, ensuring their reliable operation has become crucial nowadays [81]. For that reason, this work aims to *contribute* by evaluating the soft error reliability aspects that can impact such kind of systems.

2.2   Radiation Environment an its Effects on Semiconductors Devices

Current challenges presented in the semiconductor industry target not only performance but also the system reliability, since the capacity of transistor integration in existing technology nodes has been presenting significant issues provided by highly energetic radiation particles [82, 83]. Such particles may induce the occurrence of Single Event Effects (SEE)s on integrated transistor circuits causing the change of data states (e.g., from 0 to 1), which might incur into catastrophic results (i.e., human life losses). In order to comprehend the origin of those effects, this Section presents a brief summary of radiation effects.

Figure 2.13 illustrates the possible sources of atmospheric radiation known in the literature [84, 85, 87]. The main sources of atmospheric radiation are *galactic* and *solar* cosmic rays, which can generate high energy particles according to the particle's trajectory in the Earth's atmosphere, such as: *neutrons*, *protons* and *heavy-ions*. In addition to spacecrafts that suffer with radiation effects, such particles are those that can penetrate the planetary magnetic field, causing SEEs on devices inside the Earth's atmosphere [84].

Neutrons (Fig. 2.13A) are the primary particles generated by the shock between radiation ions and atoms from the atmosphere [88]. For that reason, such particles are more likely to generate SEEs at high altitudes. The shock of primary cosmic rays with the atoms in the air can also result in *protons*, *pions*, *kaons*, and *electrons* particles. In turn, most studies consider only *Protons* (Fig. 2.13B), since this particle have similar occurrence to *neutrons*, including energy, charging capacity and altitude. Furthermore, the *heavy-ion* particles (Fig. 2.13C) are composed by one or more electric charge units with a mass that exceeds the alpha particle. The occurrence of these particles is related to the outer layer of the Earth's atmosphere, or in the combination of high altitudes and high latitudes (polar), where they manage to cross the magnetic belt. Such effect occurs due to the interactions of heavy-ions with the Eath's atmosphere cause fragmentation and thereby remove such ions.

Once one of these ions hits a device, it can generate consequences which in turn can be classified according to the resulting effect. Figure 2.14 ilustrates the classification of radiation effects in microelectronics devices. When the equipment suffers from cumulative radiation damage and the gradual degradation such devices, known as Cumulative Effects. In other cases, radiation damage is caused by a high-energy burst, known as SEEs. Thus, aiming at elucidating the investigation promoted in this work, the next Section describes the SEE types and how they are commonly classified.

2.2.1   Classification of SEEs

This Section aims to review the types of radiation effects that lead to errors occurring in transistor-based devices. The possible types of SEEs are destructive, those that imply in hard errors mostly non recoverable, and nondestructive, soft errors that can be observed in

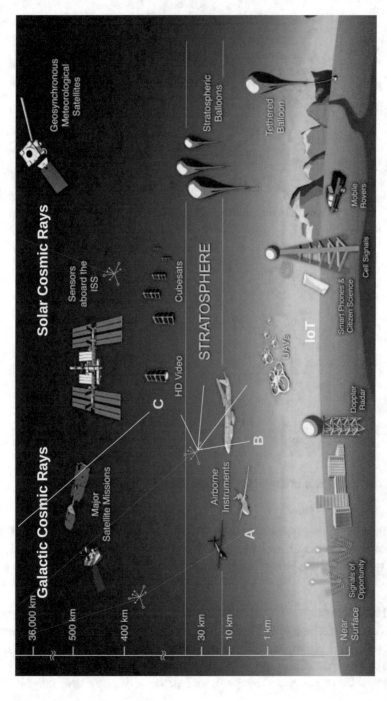

Fig. 2.13 Sketch of the Earth's radiation environment. Adapted from [84–86]

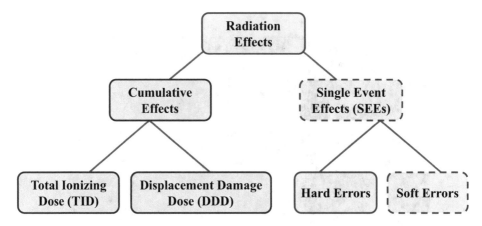

Fig. 2.14 The classification of Radiation effects. Adapted from [86, 89]

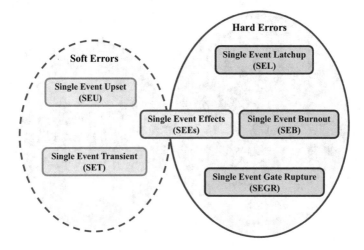

Fig. 2.15 The classification of Single Event Effects. Adapted from [86, 89, 90]

microelectronic devices (digital or analog). Soft errors imply in single particle strikes that generate some noise or crash in common workflow. Figure 2.15 illustrates the classification of SEE's types, which can be hard errors and soft errors.

2.2.1.1 Hard Errors

The hard errors comprise irreversible changes in a device, which may imply in permanent damage even after a restart. Most of the known hard errors are caused by thermal stress generated by the impact of high energy ions [91]. The aerospace and microelectronic industry have identified the occurrence of Single Event Latchup (SEL) on a wide range of devices based on Complementary Metal-Oxide-Semiconductor (CMOS) technology over the last 35 years [92–95]. With the technology scaling, the decreased transistor size along with reduced nodal capacitance increase the device sensibility to charges induced by radiation particles that can modify the electrical field and create a latchup [96, 97]. The SEL consists of charges deposited by striking energetic particles that modify the transistor electrical field and can trigger a parasitic bipolar structure, leading to device destruction as a consequence of thermal stress. In order to reduce the SEL sensitivity, the major existing solutions are based on layout and/or process optimizations. As an alternative, since latchup is known to be temperature dependent, a thermal controller that turns off the device power supply can disable the parasitic structure in case of stress, avoiding the device damage (i.e., turning this effect into a soft error).

Similarly, several studies over the past 30 years have shown that heavy ions can trigger catastrophic failure modes in power Metal-Oxide-Semiconductor Field-Effect Transistor (MOSFET)s [98, 99] Power MOSFETs, subjected to cosmic environments or with widespread occurrence of particles, are prone to Single Event Gate Rupture (SEGR) and Single Event Burnout (SEB), which can adversely affect a device's performance and can cause system failures [100]. In general, an SEB occurs in n-channel MOSFETs, combining the *off* state and abnormal applied voltages. In this situation, when an particle collides with the circuit it forms a parasitic structure that causes a high-density current and consequently the device's transistor burn [101]. Similar to SEB, SEGR effects occur mainly in MOSFETs. In this effect, the impact of accumulated ions holes under the gate, increasing the electric field through the silicon dioxide that cause dielectric break that results in a leakage current, which can also cause a door oxide's thermal failure [102]. Furthermore, the literature shows, both experimentally and theoretically, that such effects must be considered at early circuit design phase in order to reduce/avoid their occurrence [99, 103, 104].

2.2.1.2 Soft Errors

On the contrary to hard errors, soft errors are non-destructive effects induced by the impact of energetic ions. Soft errors may emerge either during intensive or idle working periods, affecting both critical and non-essential system functionalities. For instance, Processor-based systems working at sea level are expected to experience at least one soft error per day [105]. Figure 2.16 shows the occurrence of both Single Event Transient (SET) and Single Event Upset (SEU) in a given circuit example.

The difference between SET and SEU regards to the affected target and its duration over time. The SET is a momentary voltage spike at a node of an integrated circuit generated by a single energetic particle passing through or near the junctions creating an ionization track. As illustrated in Fig. 2.16, the peak voltage affects non-latched elements such as combinational logic [107]. The resulting effect may propagate any significant distance through the combinational logic and, if it is not masked, can reach a memory element where can be sampled causing a fault [108]. In this sense, an SET can become an SEU when it reaches

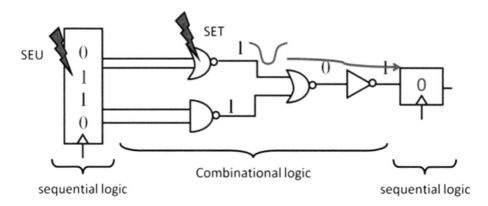

Fig. 2.16 SEU and SET effects occurring in a circuit example from [106]

an memory element, since it is stored in a memory element the error is no longer transient. Although these effects are recurrent and widely studied, other radiation effects appear with reduced scaling in current technology nodes. For instance, the charge sharing causes signal degradation by transferring charges from one electronic domain to another [109]. Furthermore, the reduced scale CMOS circuits have been showing a significant sensitivity to SETs [110, 111]. Despite that, recent technology nodes consider either Fin Field Effect Transistor (FinFET) and Fully Depleted Silicon On Insulator (FDSOI) nanotechnologies that show significant improvements in terms of fault resilience [112, 113].

As shown in Fig. 2.16, the SEU differs from SET with regard to the affected target in the integrated circuit and its duration over the time. The SEU phenomenon may occur in a digital circuit when an ionizing particle strike results to a change in data states from a storage element such as latches, flip-flops, Random Access Memory (RAM) cells, or asynchronous memory logic [114]. When a SEU occurs, a simple reset or rewriting masks the resultant soft error effect, preserving device's normal behavior. In this sense, logical masking can occur through an operation that instantly overwrites the faulty value in the memory element. Besides that, an electrical masking can occur when the signal interference is not enough to change the logic state. However, in the worst case scenario, the SEU effect can remain several clock cycles, for synchronous logic, or until the next transition of an input signal in asynchronous logic [115]. In this case, the soft error could imply in catastrophic results in the circuit operation.

References

1. Mahdavinejad, M.S., Rezvan, M., Barekatain, M., Adibi, P., Barnaghi, P., Sheth, A.P.: Machine learning for internet of things data analysis: a survey. Digit. Commun. Netw. **4**(3), 161–175 (2018). https://doi.org/10.1016/j.dcan.2017.10.002

2. Samuel, A.L.: Some studies in machine learning using the game of checkers. IBM J. Res. Dev. **3**(3), 210–229 (1959). https://doi.org/10.1147/rd.33.0210

3. Ayodele, T.O.: Types of machine learning algorithms. New Adv. Mach. Learn. **3**, 19–48 (2010). https://doi.org/10.5772/9385

4. Kato, N., Fadlullah, Z.M., Mao, B., Tang, F., Akashi, O., Inoue, T., Mizutani, K.: The deep learning vision for heterogeneous network traffic control: proposal, challenges, and future perspective. IEEE Wirel. Commun. **24**(3), 146–153 (2017). https://doi.org/10.1109/MWC.2016.1600317WC

5. Shanthamallu, U.S., Spanias, A., Tepedelenlioglu, C., Stanley, M.: A brief survey of machine learning methods and their sensor and iot applications. In: 2017 8th International Conference on Information, Intelligence, Systems Applications (IISA), pp. 1–8 (2017). https://doi.org/10.1109/IISA.2017.8316459

6. Chen, J., Ran, X.: Deep learning with edge computing: a review. Proc. IEEE **107**(8), 1655–1674 (2019). https://doi.org/10.1109/JPROC.2019.2921977

7. Marchisio, A., Hanif, M.A., Khalid, F., Plastiras, G., Kyrkou, C., Theocharides, T., Shafique, M.: Deep learning for edge computing: current trends, cross-layer optimizations, and open research challenges. In: 2019 IEEE Computer Society Annual Symposium on VLSI (ISVLSI), pp. 553–559 (2019). https://doi.org/10.1109/ISVLSI.2019.00105

8. Barlow, H.B.: Unsupervised learning. Neural Comput. **1**(3), 295–311 (1989). https://doi.org/10.1162/neco.1989.1.3.295

9. Hastie, T., Tibshirani, R., Friedman, J.: Unsupervised learning. In: The Elements of Statistical Learning, pp. 485–585. Springer (2009). https://doi.org/10.1007/978-0-387-84858-7

10. Celebi, M.E., Aydin, K.: Unsupervised Learning Algorithms. Springer (2016). https://doi.org/10.1007/978-3-319-24211-8

11. Kanungo, T., Mount, D.M., Netanyahu, N.S., Piatko, C.D., Silverman, R., Wu, A.Y.: An efficient k-means clustering algorithm: analysis and implementation. IEEE Trans. Pattern Anal. Mach. Intell. **24**(7), 881–892 (2002). https://doi.org/10.1109/TPAMI.2002.1017616

12. Sonagara, D., Badheka, S.: Comparison of basic clustering algorithms. Int. J. Comput. Sci. Mob. Comput. **3**(10), 58–61 (2014). https://www.ijcsmc.com/docs/papers/October2014/V3I10201405.pdf

13. Hathaway, R.J., Bezdek, J.C.: Extending fuzzy and probabilistic clustering to very large data sets. Comput. Stat. Data Anal. **51**(1), 215–234 (2006). https://doi.org/10.1016/j.csda.2006.02.008. The Fuzzy Approach to Statistical Analysis

14. Ghosh, S., Dubey, S.K.: Comparative analysis of k-means and fuzzy c-means algorithms. Int. J. Adv. Comput. Sci. Appl. **4**(4) (2013). https://doi.org/10.14569/IJACSA.2013.040406

15. Figueiredo, M.A.T., Jain, A.K.: Unsupervised learning of finite mixture models. IEEE Trans. Pattern Anal. Mach. Intell. **24**(3), 381–396 (2002). https://doi.org/10.1109/34.990138

16. Xu, H., Li, G.: Density-based probabilistic clustering of uncertain data. In: 2008 International Conference on Computer Science and Software Engineering, vol. 4, pp. 474–477. IEEE (2008). https://doi.org/10.1109/CSSE.2008.968

17. Weinberger, K.Q., Sha, F., Saul, L.K.: Learning a kernel matrix for nonlinear dimensionality reduction. In: Proceedings of the Twenty-First International Conference on Machine Learning, ICML '04, p. 106. Association for Computing Machinery, New York, NY, USA (2004). https://doi.org/10.1145/1015330.1015345

18. Van Der Maaten, L., Postma, E., Van den Herik, J.: Dimensionality reduction: a comparative review. J. Mach. Learn. Res.—JMLR **10**(66–71), 13 (2009). https://www.researchgate.net/publication/228657549_Dimensionality_Reduction_A_Comparative_Review

19. Khalid, S., Khalil, T., Nasreen, S.: A survey of feature selection and feature extraction techniques in machine learning. In: 2014 Science and Information Conference, pp. 372–378. IEEE (2014). https://doi.org/10.1109/SAI.2014.6918213

20. Dy, J.G., Brodley, C.E.: Feature selection for unsupervised learning. J. Mach. Learn. Res. **5**(Aug), 845–889 (2004). https://dl.acm.org/doi/10.5555/1005332.1016787

21. Hild, K.E., Erdogmus, D., Torkkola, K., Principe, J.C.: Feature extraction using information-theoretic learning. IEEE Trans. Pattern Anal. Mach. Intell. **28**(9), 1385–1392 (2006). https://doi.org/10.1109/TPAMI.2006.186

22. Tsuge, S., Shishibori, M., Kuroiwa, S., Kita, K.: Dimensionality reduction using non-negative matrix factorization for information retrieval. In: 2001 IEEE International Conference on Systems, Man and Cybernetics. e-Systems and e-Man for Cybernetics in Cyberspace (Cat.No.01CH37236), vol. 2, pp. 960–965 (2001). https://doi.org/10.1109/ICSMC.2001.973042

23. Smyth, P., Wolpert, D.: Stacked density estimation. Adv. Neural Inf. Process. Syst. **10**, 668–674 (1997). https://dl.acm.org/doi/10.5555/3008904.3008999

24. Trentin, E., Freno, A.: Unsupervised nonparametric density estimation: a neural network approach. In: 2009 International Joint Conference on Neural Networks, pp. 3140–3147. IEEE (2009). https://doi.org/10.1109/IJCNN.2009.5179010

25. Dinh, L., Sohl-Dickstein, J., Bengio, S.: Density Estimation Using Real NVP (2016). arXiv:1605.08803

26. Pedregosa, F., Varoquaux, G., Gramfort, A., Michel, V., Thirion, B., Grisel, O., Blondel, M., Prettenhofer, P., Weiss, R., Dubourg, V., et al.: Scikit-learn: machine learning in python. J. Mach. Learn. Res. **12**, 2825–2830 (2011). https://scikit-learn.org/

27. Gudivada, V., Apon, A., Ding, J.: Data quality considerations for big data and machine learning: Going beyond data cleaning and transformations. Int. J. Adv. Soft. **10**(1), 1–20 (2017). https://www.researchgate.net/publication/318432363_Data_Quality_Considerations_for_Big_Data_and_Machine_Learning_Going_Beyond_Data_Cleaning_and_Transformations

28. Caruana, R., Niculescu-Mizil, A.: An empirical comparison of supervised learning algorithms. In: Proceedings of the 23rd International Conference on Machine Learning, ICML '06, pp. 161–168. Association for Computing Machinery, New York, NY, USA (2006). https://doi.org/10.1145/1143844.1143865

29. Morente-Molinera, J.A., Mezei, J., Carlsson, C., Herrera-Viedma, E.: Improving supervised learning classification methods using multigranular linguistic modeling and fuzzy entropy. IEEE Trans. Fuzzy Syst. **25**(5), 1078–1089 (2017). https://doi.org/10.1109/TFUZZ.2016.2594275

30. Xie, S., Liu, Y.: Improving supervised learning for meeting summarization using sampling and regression. Comput. Speech Lang. **24**(3), 495–514 (2010). https://doi.org/10.1016/j.csl.2009.04.007

31. Jain, V., Murray, J.F., Roth, F., Turaga, S., Zhigulin, V., Briggman, K.L., Helmstaedter, M.N., Denk, W., Seung, H.S.: Supervised learning of image restoration with convolutional networks. In: 2007 IEEE 11th International Conference on Computer Vision, pp. 1–8 (2007). https://doi.org/10.1109/ICCV.2007.4408909

32. Valentini, G., Dietterich, T.G.: Bias-variance analysis of support vector machines for the development of SVM-based ensemble methods. J. Mach. Learn. Res. **5**(Jul), 725–775 (2004). https://www.jmlr.org/papers/volume5/valentini04a/valentini04a.pdf

33. Belkin, M., Hsu, D., Ma, S., Mandal, S.: Reconciling modern machine-learning practice and the classical bias–variance trade-off. Proc. Natl. Acad. Sci. **116**(32), 15849–15854 (2019). https://doi.org/10.1073/pnas.1903070116

34. Reppen, A.M., Soner, H.M.: Bias-variance trade-off and overlearning in dynamic decision problems (2020). arXiv:2011.09349
35. Schmidhuber, J.: Deep learning in neural networks: an overview. Neural Netw. **61**, 85–117 (2015). https://doi.org/10.1016/j.neunet.2014.09.003
36. Lee, W., Kim, S., Lee, Y.T., Lee, H.W., Choi, M.: Deep neural networks for wild fire detection with unmanned aerial vehicle. In: 2017 IEEE International Conference on Consumer Electronics (ICCE), pp. 252–253 (2017). https://doi.org/10.1109/ICCE.2017.7889305
37. Gibril, M.B.A., Shafri, H.Z.M., Shanableh, A., Al-Ruzouq, R., Wayayok, A., Hashim, S.J.: Deep convolutional neural network for large-scale date palm tree mapping from uav-based images. Remote Sens. **13**(14) (2021). https://doi.org/10.3390/rs13142787
38. Nugraha, B.T., Su, S.F., Fahmizal: Towards self-driving car using convolutional neural network and road lane detector. In: 2017 2nd International Conference on Automation, Cognitive Science, Optics, Micro Electro-Mechanical System, and Information Technology (ICACOMIT), pp. 65–69 (2017). https://doi.org/10.1109/ICACOMIT.2017.8253388
39. Rao, Q., Frtunikj, J.: Deep learning for self-driving cars: chances and challenges. In: Proceedings of the 1st International Workshop on Software Engineering for AI in Autonomous Systems, SEFAIS '18, pp. 35–38. Association for Computing Machinery, New York, NY, USA (2018). https://doi.org/10.1145/3194085.3194087
40. Abdel-Hamid, O., Mohamed, A.r., Jiang, H., Deng, L., Penn, G., Yu, D.: Convolutional neural networks for speech recognition. IEEE/ACM Trans. Audio Speech Lang. Process. **22**(10), 1533–1545 (2014). https://doi.org/10.1109/TASLP.2014.2339736
41. Nassif, A.B., Shahin, I., Attili, I., Azzeh, M., Shaalan, K.: Speech recognition using deep neural networks: a systematic review. IEEE Access **7**, 19143–19165 (2019). https://doi.org/10.1109/ACCESS.2019.2896880
42. Wei, J., Han, J., Cao, S.: Satellite iot edge intelligent computing: a research on architecture. Electronics **8**(11) (2019). https://doi.org/10.3390/electronics8111247
43. Sudharsan, B., Breslin, J.G., Ali, M.I.: Rce-nn: A five-stage pipeline to execute neural networks (cnns) on resource-constrained iot edge devices. In: Proceedings of the 10th International Conference on the Internet of Things, IoT '20. Association for Computing Machinery, New York, NY, USA (2020). https://doi.org/10.1145/3410992.3411005
44. Wang, S.C.: Artificial Neural Network, pp. 81–100. Springer US, Boston, MA (2003). https://doi.org/10.1007/978-1-4615-0377-4_5
45. Guresen, E., Kayakutlu, G.: Definition of artificial neural networks with comparison to other networks. Procedia Comput. Sci. **3**, 426–433 (2011). https://doi.org/10.1016/j.procs.2010.12.071. World Conference on Information Technology
46. Haykin, S.: Neural Networks: A Comprehensive Foundation, 3rd edn. Prentice-Hall Inc., USA (2007). https://dl.acm.org/doi/book/10.5555/1213811
47. Wang, J.: Analysis and design of a recurrent neural network for linear programming. IEEE Trans. Circuits Syst. I: Fund. Theory Appl. **40**(9), 613–618 (1993). https://doi.org/10.1109/81.244913
48. Bruguier, A., Gnanapragasam, D., Johnson, L., Rao, K., Beaufays, F.: Pronunciation learning with rnn-transducers. In: Proceedings of Interspeech 2017, pp. 2556–2560 (2017). https://doi.org/10.21437/Interspeech.2017-47
49. Sigtia, S., Haynes, R., Richards, H., Marchi, E., Bridle, J.: Efficient voice trigger detection for low resource hardware. In: Interspeech, pp. 2092–2096 (2018). https://doi.org/10.21437/Interspeech.2018-2204
50. Goodfellow, I., Bengio, Y., Courville, A.: Sequence modeling: recurrent and recursive nets. Deep Learn. 367–415 (2016). http://www.deeplearningbook.org

51. Fan, Y., Lu, X., Li, D., Liu, Y.: Video-based emotion recognition using CNN-RNN and c3d hybrid networks. In: Proceedings of the 18th ACM International Conference on Multimodal Interaction, ICMI '16, pp. 445–450. Association for Computing Machinery, New York, NY, USA (2016). https://doi.org/10.1145/2993148.2997632

52. Liu, Q., Zhang, N., Yang, W., Wang, S., Cui, Z., Chen, X., Chen, L.: A review of image recognition with deep convolutional neural network. In: Huang, D.S., Bevilacqua, V., Premaratne, P., Gupta, P. (eds.), Intelligent Computing Theories and Application, pp. 69–80. Springer International Publishing, Cham (2017). https://doi.org/10.1007/978-3-319-63309-1_7

53. Ullah, A., Ahmad, J., Muhammad, K., Sajjad, M., Baik, S.W.: Action recognition in video sequences using deep bi-directional lstm with CNN features. IEEE Access **6**, 1155–1166 (2018). https://doi.org/10.1109/ACCESS.2017.2778011

54. Van den Oord, A., Dieleman, S., Schrauwen, B.: Deep content-based music recommendation. Adv. Neural Inf. Process. Syst. **26** (2013). https://dl.acm.org/doi/10.5555/2999792.2999907

55. Kim, D., Park, C., Oh, J., Lee, S., Yu, H.: Convolutional matrix factorization for document context-aware recommendation. In: Proceedings of the 10th ACM Conference on Recommender Systems, pp. 233–240 (2016). https://doi.org/10.1145/2959100.2959165

56. Seo, S., Huang, J., Yang, H., Liu, Y.: Interpretable convolutional neural networks with dual local and global attention for review rating prediction. In: Proceedings of the Eleventh ACM Conference on Recommender Systems, pp. 297–305 (2017). https://doi.org/10.1145/3109859.3109890

57. Zhang, S., Yao, L., Sun, A., Tay, Y.: Deep learning based recommender system: a survey and new perspectives. ACM Comput. Surv. (CSUR) **52**(1), 1–38 (2019). https://doi.org/10.1145/3285029

58. Krizhevsky, A., Hinton, G., et al.: CIFAR-10/100—Learning multiple layers of features from tiny images. Technical report, University of Toronto (2009). http://citeseerx.ist.psu.edu/viewdoc/download?doi=10.1.1.222.9220&rep=rep1&type=pdf

59. Li, Q., Cai, W., Wang, X., Zhou, Y., Feng, D.D., Chen, M.: Medical image classification with convolutional neural network. In: 2014 13th International Conference on Control Automation Robotics & Vision (ICARCV), pp. 844–848. IEEE (2014). https://doi.org/10.1109/ICARCV.2014.7064414

60. Wang, J., Yang, Y., Mao, J., Huang, Z., Huang, C., Xu, W.: CNN-RNN: a unified framework for multi-label image classification. In: Proceedings of the IEEE Conference on Computer Vision and Pattern Recognition, pp. 2285–2294 (2016). https://doi.ieeecomputersociety.org/10.1109/CVPR.2016.251

61. Howard, A.G., Zhu, M., Chen, B., Kalenichenko, D., Wang, W., Weyand, T., Andreetto, M., Adam, H.: MobileNets: Efficient Convolutional Neural Networks for Mobile Vision Applications (2017). arXiv:1704.04861

62. Lee, H., Kwon, H.: Going deeper with contextual cnn for hyperspectral image classification. IEEE Trans. Image Process. **26**(10), 4843–4855 (2017). https://doi.org/10.1109/TIP.2017.2725580

63. Hussain, M., Bird, J.J., Faria, D.R.: A study on CNN transfer learning for image classification. In: UK Workshop on computational Intelligence, pp. 191–202. Springer (2018). https://doi.org/10.1007/978-3-319-97982-3_16

64. Zhang, M., Li, W., Du, Q.: Diverse region-based CNN for hyperspectral image classification. IEEE Trans. Image Process. **27**(6), 2623–2634 (2018). https://doi.org/10.1109/TIP.2018.2809606

65. Albawi, S., Mohammed, T.A., Al-Zawi, S.: Understanding of a convolutional neural network. In: 2017 International Conference on Engineering and Technology (ICET), pp. 1–6. IEEE (2017). https://doi.org/10.1109/ICEngTechnol.2017.8308186

66. Hermann, M., Hake, F., Alkhatib, H., Hesse, C., Holste, K., Umlauf, G., Kermarrec, G., Neumann, I.: Damage detection for port infrastructure by means of machine-learning-algorithms. In: Smart surveyors for land and water management, FIG Working Week 2020, Amsterdam, the Netherlands, 10–14 May 2020. International Federation of Surveyors, FIG (2020). https://opus.htwg-konstanz.de/frontdoor/index/index/docId/2630

67. Szydlo, T., Sendorek, J., Brzoza-Woch, R.: Enabling machine learning on resource constrained devices by source code generation of the learned models. In: International Conference on Computational Science, pp. 682–694. Springer (2018). https://doi.org/10.1007/978-3-319-93701-4_54

68. Dawit, M., Frisk, F.: Edge machine learning for energy efficiency of resource constrained iot devices. In: SPWID 2019: The Fifth International Conference on Smart Portable, Wearable, Implantable and Disability oriented Devices and Systems, pp. 9–14 (2019). https://www.diva-portal.org/smash/record.jsf?pid=diva23A1462789&dswid=9780

69. Qi, X., Liu, C.: Enabling deep learning on iot edge: approaches and evaluation. In: 2018 IEEE/ACM Symposium on Edge Computing (SEC), pp. 367–372 (2018). https://doi.org/10.1109/SEC.2018.00047

70. Sun, D., Liu, S., Gaudiot, J.L.: Enabling Embedded Inference Engine with Arm Compute Library: a Case Study (2017). arXiv:1704.03751

71. Lai, L., Suda, N., Chandra, V.: CMSIS-NN: Efficient Neural Network Kernels for Arm Cortex-M Cpus (2018). arXiv:1801.06601

72. Garofalo, A., Rusci, M., Conti, F., Rossi, D., Benini, L.: Pulp-nn: a computing library for quantized neural network inference at the edge on risc-v based parallel ultra low power clusters. In: 2019 26th IEEE International Conference on Electronics, Circuits and Systems (ICECS), pp. 33–36 (2019). https://doi.org/10.1109/ICECS46596.2019.8965067

73. STMicroelectronics: AI expansion pack for STM32CubeMX (2020). https://www.st.com/en/embedded-software/x-cube-ai.html

74. Capotondi, A., Rusci, M., Fariselli, M., Benini, L.: Cmix-nn: Mixed low-precision CNN library for memory-constrained edge devices. IEEE Trans. Circuits Syst. II: Express Briefs 67(5), 871–875 (2020). https://doi.org/10.1109/TCSII.2020.2983648

75. Amoh, J., Odame, K.M.: An optimized recurrent unit for ultra-low-power keyword spotting. ACM Interact. Mob. Wearable Ubiquitous Technol. 3(2) (2019). https://doi.org/10.1145/3328907

76. Han, S., Mao, H., Dally, W.J.: Deep Compression: Compressing Deep Neural Networks with Pruning, Trained Quantization and Huffman Coding (2015). arXiv:1510.00149

77. He, Y., Zhang, X., Sun, J.: Channel pruning for accelerating very deep neural networks. In: Proceedings of the IEEE International Conference on Computer Vision, pp. 1389–1397 (2017). https://doi.ieeecomputersociety.org/10.1109/ICCV.2017.155

78. Wu, H., Judd, P., Zhang, X., Isaev, M., Micikevicius, P.: Integer Quantization for Deep Learning Inference: Principles and Empirical Evaluation (2020). arXiv:2004.09602. https://doi.org/10.48550/arXiv.2004.09602

79. Gupta, S., Agrawal, A., Gopalakrishnan, K., Narayanan, P.: Deep learning with limited numerical precision. In: F. Bach, D. Blei (eds.) Proceedings of the 32nd International Conference on Machine Learning, Proceedings of Machine Learning Research, vol. 37, pp. 1737–1746. Lille, France (2015). https://dl.acm.org/doi/abs/10.5555/3045118.3045303

80. Judd, P., Albericio, J., Hetherington, T., Aamodt, T., Jerger, N.E., Urtasun, R., Moshovos, A.: Reduced-Precision Strategies for Bounded Memory in Deep Neural Nets (2015). arXiv:1511.05236. https://doi.org/10.48550/arXiv.1511.05236

81. Mittal, S.: A survey on modeling and improving reliability of DNN algorithms and accelerators. J. Syst. Arch. 104, 101689 (2020). https://doi.org/10.1016/j.sysarc.2019.101689

82. Mansour, W., Velazco, R.: SEU fault-injection in vhdl-based processors: a case study. J. Electron. Test. **29**(1), 87–94 (2013). https://doi.org/10.1007/s10836-013-5351-6

83. ISO: Road vehicles–Functional safety (2011). https://www.iso.org/standard/68383.html

84. Barth, J., Dyer, C., Stassinopoulos, E.: Space, atmospheric, and terrestrial radiation environments. IEEE Trans. Nucl. Sci. **50**(3), 466–482 (2003). https://doi.org/10.1109/TNS.2003.813131

85. Ciani, L., Catelani, M., Veltroni, L.: Fault tolerant techniques to diagnose and mitigate single event upset (seu) effects on electronic programmable devices. In: Proceedings of 16th ImEKo TC4 Symposium. Citeseer (2008). https://www.imeko.org/publications/tc4-2008/IMEKO-TC4-2008-216.pdf

86. Mutuel, L.H.: Single event effects mitigation techniques report. Technical report, Federal Aviation Administration, William J. Hughes Technical Center (2016). https://www.faa.gov/aircraft/air_cert/design_approvals/air_software/media/TC-15-62.pdf

87. Ziegler, J.F.: Terrestrial cosmic rays. IBM J. Res. Dev. **40**(1), 19–39 (1996). https://doi.org/10.1147/rd.401.0019

88. Normand, E.: Single-event effects in avionics. IEEE Trans. Nucl. Sci. **43**(2), 461–474 (1996). https://doi.org/10.1109/23.490893

89. Stassinopoulos, E., Raymond, J.: The space radiation environment for electronics. Proc. IEEE **76**(11), 1423–1442 (1988). https://doi.org/10.1109/5.90113

90. Kobayashi, D.: Scaling trends of digital single-event effects: a survey of seu and set parameters and comparison with transistor performance. IEEE Trans. Nucl. Sci. **68**(2), 124–148 (2021). https://doi.org/10.1109/TNS.2020.3044659

91. Gomez Toro, D., Seguin, F., Arzel, M., Jézéquel, M.: Study of a cosmic ray impact on combinatorial logic circuits of an 8bit SAR ADC in 65 nm CMOS technology. In: 2013 IEEE 56th International Midwest Symposium on Circuits and Systems (MWSCAS), pp. 241–244 (2013). https://doi.org/10.1109/MWSCAS.2013.6674630

92. Soliman, K., Nichols, D.K.: Latchup in cmos devices from heavy ions. IEEE Trans. Nucl. Sci. **30**(6), 4514–4519 (1983). https://doi.org/10.1109/TNS.1983.4333163

93. Johnston, A.: The influence of vlsi technology evolution on radiation-induced latchup in space systems. IEEE Trans. Nucl. Sci. **43**(2), 505–521 (1996). https://doi.org/10.1109/23.490897

94. Bruguier, G., Palau, J.M.: Single particle-induced latchup. IEEE Trans. Nucl. Sci. **43**(2), 522–532 (1996). https://doi.org/10.1109/23.490898

95. Hutson, J., Pellish, J., Tipton, A., Boselli, G., Xapsos, M., Kim, H., Friendlich, M., Campola, M., Seidleck, S., LaBel, K., et al.: Evidence for lateral angle effect on single-event latchup in 65 nm SRAMS. IEEE Trans. Nucl. Sci. **56**(1), 208–213 (2009). https://doi.org/10.1109/TNS.2008.2010395

96. Schwank, J., Shaneyfelt, M., Baggio, J., Dodd, P., Felix, J., Ferlet-Cavrois, V., Paillet, P., Lambert, D., Sexton, F., Hash, G., et al.: Effects of particle energy on proton-induced single-event latchup. IEEE Trans. Nucl. Sci. **52**(6), 2622–2629 (2005). https://doi.org/10.1109/TNS.2005.860672

97. Schwank, J.R., Shaneyfelt, M.R., Baggio, J., Dodd, P., Felix, J., Ferlet-Cavrois, V., Paillet, P., Lum, G., Girard, S., Blackmore, E.: Effects of angle of incidence on proton and neutron-induced single-event latchup. IEEE Trans. Nucl. Sci. **53**(6), 3122–3131 (2006). https://doi.org/10.1109/TNS.2006.884059

98. Johnson, G., Palau, J., Dachs, C., Galloway, K., Schrimpf, R.: A review of the techniques used for modeling single-event effects in power mosfets. IEEE Trans. Nucl. Sci. **43**(2), 546–560 (1996). https://doi.org/10.1109/23.490900

99. Titus, J.L.: An updated perspective of single event gate rupture and single event burnout in power mosfets. IEEE Trans. Nucl. Sci. **60**(3), 1912–1928 (2013). https://doi.org/10.1109/TNS. 2013.2252194

100. Hands, A., Morris, P., Ryden, K., Dyer, C., Truscott, P., Chugg, A., Parker, S.: Single event effects in power mosfets due to atmospheric and thermal neutrons. IEEE Trans. Nucl. Sci. **58**(6), 2687–2694 (2011). https://doi.org/10.1109/TNS.2011.2168540

101. Wrobel, T.F., Coppage, F.N., Hash, G.L., Smith, A.J.: Current induced avalanche in epitaxial structures. IEEE Trans. Nucl. Sci. **32**(6), 3991–3995 (1985). https://doi.org/10.1109/TNS.1985. 4334056

102. Titus, J., Wheatley, C., Van Tyne, K., Krieg, J., Burton, D., Campbell, A.: Effect of ion energy upon dielectric breakdown of the capacitor response in vertical power mosfets. IEEE Trans. Nucl. Sci. **45**(6), 2492–2499 (1998). https://doi.org/10.1109/23.736490

103. Barak, J., Haran, A., David, D., Rapaport, S.: A double-power-mosfet circuit for protection from single event burnout. IEEE Trans. Nucl. Sci. **55**(6), 3467–3472 (2008). https://doi.org/10. 1109/TNS.2008.2007486

104. Mutuel, L.H.: Appreciating the effectiveness of single event effect mitigation techniques. In: 2014 IEEE/AIAA 33rd Digital Avionics Systems Conference (DASC), pp. 5B1–1. IEEE (2014). https://doi.org/10.1109/DASC.2014.6979481

105. Granlund, T., Granbom, B., Olsson, N.: Soft error rate increase for new generations of SRAMS. IEEE Trans. Nucl. Sci. **50**(6), 2065–2068 (2003). https://doi.org/10.1109/TNS.2003.821593

106. Kastensmidt, F., Rech, P.: Radiation effects and fault tolerance techniques for FPGAS and GPUS. In: FPGAs and Parallel Architectures for Aerospace Applications, pp. 3–17. Springer (2016). https://doi.org/10.1007/978-3-319-14352-1_1

107. Loveless, T., Kauppila, J., Jagannathan, S., Ball, D., Rowe, J., Gaspard, N., Atkinson, N., Blaine, R., Reece, T., Ahlbin, J., et al.: On-chip measurement of single-event transients in a 45 nm silicon-on-insulator technology. IEEE Trans. Nucl. Sci. **59**(6), 2748–2755 (2012). https:// doi.org/10.1109/TNS.2012.2218257

108. Wirth, G., Kastensmidt, F.L., Ribeiro, I.: Single event transients in logic circuits-load and propagation induced pulse broadening. IEEE Trans. Nucl. Sci. **55**(6), 2928–2935 (2008). https:// doi.org/10.1109/TNS.2008.2006265

109. Ferlet-Cavrois, V., Massengill, L.W., Gouker, P.: Single event transients in digital CMOS—a review. IEEE Trans. Nucl. Sci. **60**(3), 1767–1790 (2013). https://doi.org/10.1109/TNS.2013. 2255624

110. Dodd, P., Shaneyfelt, M., Schwank, J., Felix, J.: Current and future challenges in radiation effects on CMOS electronics. IEEE Trans. Nucl. Sci. **57**(4), 1747–1763 (2010). https://doi.org/ 10.1109/TNS.2010.2042613

111. Prinzie, J., Steyaert, M., Leroux, P.: Radiation effects in CMOS technology. In: Radiation Hardened CMOS Integrated Circuits for Time-Based Signal Processing, pp. 1–20. Springer (2018). https://doi.org/10.1007/978-3-319-78616-2_1

112. Liu, R., Evans, A., Chen, L., Li, Y., Glorieux, M., Wong, R., Wen, S.J., Cunha, J., Summerer, L., Ferlet-Cavrois, V.: Single event transient and TID study in 28 nm UTBB FDSOI technology. IEEE Trans. Nucl. Sci. **64**(1), 113–118 (2016). https://doi.org/10.1109/TNS.2016.2627015

113. de Aguiar, Y., Artola, L., Hubert, G., Meinhardt, C., Kastensmidt, F.L., Reis, R.: Evaluation of radiation-induced soft error in majority voters designed in 7 nm finfet technology. Microelectron. Reliab. **76**, 660–664 (2017). https://doi.org/10.1016/j.microrel.2017.06.077

114. Vargas, F., Nicolaidis, M.: Seu-tolerant SRAM design based on current monitoring. In: Proceedings of IEEE 24th International Symposium on Fault-Tolerant Computing, pp. 106–115. IEEE (1994). https://doi.org/10.1109/FTCS.1994.315652
115. Taber, A., Normand, E.: Single event upset in avionics. IEEE Trans. Nucl. Sci. **40**(2), 120–126 (1993). https://doi.org/10.1109/23.212327

Related Works

The soft error assessment and mitigation literature is abundant, requiring a taxonomy to classify the different approaches. This book considers the definitions from Avižienis et al. [1] for fault, error, and failure. A fault is an event that may cause the internal state of the system to change, e.g., a radiation particle strike. When a fault affects the system's internal state, it becomes an error. If the error causes a deviation of at least one of the system's external states, then it is considered as a failure. To achieve compliance with safety and reliability standard requirements, it is utmost importance to provide systems with appropriate mechanisms to tackle systematic, SEU, or SET faults, also known as soft errors.

In this regard, this Chapter presents a literature review of the works related to this Book contributions on the soft error reliability assessment of ML models executing on resource-constrained IoT systems. First, Sect. 3.1 presents a review of fault injector frameworks implemented on the top of VPs. Next, Sect. 3.2 discusses some related works on soft error reliability assessment of ML models in different scopes. Finally, we distinguish this Book from the works found in the literature (Sects. 3.2.2 and 3.1.1).

3.1   Soft Error Assesment Considering Virtual Platforms

Fault injection simulation frameworks are gaining momentum due to their efficiency to obtain prominent results on the soft error resilience of different systems early in the design phase. These FI frameworks can be divided into two classes: (*i*) those focused on the accuracy of the results and (*ii*) those that deal with highly complex systems or have little time to explore the project reliability. Simulation-based soft error analysis at RTL or gate levels are examples of the first class. Within this category, Mansour and Velazco [2] presented a method called Direct Fault Injection (DFI) to emulate the consequences of a SEU occurring in the processor's memory cells. However, this approach might, in same cases, require modification of the circuit architecture, which is not easily applied anywhere. Abbasitabar et al. [3] also

© The Author(s), under exclusive license to Springer Nature Switzerland AG 2023   41
G. Abich et al., *Early Soft Error Reliability Assessment of Convolutional Neural Networks Executing on Resource-Constrained IoT Edge Devices*, Synthesis Lectures on Engineering, Science, and Technology, https://doi.org/10.1007/978-3-031-18599-1_3

investigated the fault susceptibility of the LEON3 processor in RTL, but their approach relies on the Modelsim simulator to inject faults without explicitly changing the reference design. Although both approaches produce accurate results, they are restricted to relatively small or specific systems, where experimental fault campaigns provide some simulation performance but with low observability (i.e., fault must be detected by the system).

Researchers started investigating other approaches, such as the use of VPs, aiming to boost the soft error analysis of complex systems comprising not only real software stacks but also ISAs and state-of-the-art processors [6, 7, 13, 14]. VP simulators facilitate fault injection implementation and analyses due to their design flexibility (e.g., several processor models available) and debugging capabilities (e.g., GDB support). Tables 3.1 and 3.2 show related works considering fault injection capabilities in VP simulators (e.g., Microarchitectural Memory Component (MMC), Register File, and Memory Physical Addressing), which are detailed below.

Authors in [13] present the Relyzer tool set, a hybrid simulation framework for Scalable Processor Architecture (SPARC) core using Simics [15] and General Execution-driven Multiprocessor Simulator (GEMS) [16] simulators coupled with a pruning technique to reduce injected faults and targeting architectural integer registers and in the output latches of the address generation units. The low number of injected faults is due to the high-cost simulation time of the simics+GEMS simulator, which can achieve a few hundred Kilo Instructions Per Second (KIPS). Further Hari et al. [4] presented the GangES approach, a Relyzer extension to more aggressive pruning and reducing the number of faults that must be simulated. With these techniques embedded in the GangES, the authors further reduce this simulation time by half. In [5], the authors proposed a fault injection framework based on Quick Emulator (QEMU) [17] to inject faults in an x86 architecture running applications in a Real-Time Operating System (RTOS). This approach considers an in-house experimental setup that achieved some performance related to the number of injected faults (i.e., an average of less than one fault per second).

Parasyris et al. [6] introduced the GemFI, a fault injection tool based on the cycle-accurate full-system model of the gem5 simulator [18]. Authors improve the fault injections by creating checkpoints and parallelization strategies to gain simulation performance (i.e., 64.5x faster on average). Moreover, the authors in [7] propose two tools: the GeFIN tool, a gem5-based fault injection framework and MaFIN, a MIPS Assembler and Runtime Simulator (MARSS)-based [19] fault injection framework. In this work, faults are injected, randomly in time, in general-purpose registers, caches control registers, and other micro architectural components. Reference [14] presented a fault injection framework based on another VP, the OVPsim [20], which has the advantage of supporting parallel simulation to boost up the fault injection process. Such works denote the VP fault injection simulators not only has higher performance but the flexibility to perform techniques that reduce the simulation time (e.g., checkpoints and parallelization) considering much more complex scenarios.

To cover higher fault probabilities, the authors in [8] introduces the gemV, a gem5-based [18] fault injection framework for micro-architectural elements such as instruction queue,

Table 3.1 Related works considering virtual platform fault injection simulators

Year	Author	Simulator	ISAs	Applications	Guest OS	Level of detailing
2014	Hari et al. [4] Relyzer	Simics GEMS	SPARC V9	12: Parsec 2, Splash-2, and SPEC-Int 2006	OpenSolaris	Cycle and instruction
2014	Geissler et al. [5]	QEMU	x86	4 in-house applications	RTEMS	Instruction
2014	Parasyris et al. [6] GemFI	gem5	Alpha ISA	6 baremetal	None or Unknown	Cycle
2015	Kaliorakis et al. [7] MaFIN and GeFIN	MARSS gem5	x86 ARMv7	10 MiBench	None or Unknown	Cycle
2016	Tanikella et al. [8] gemV	gem5	x86 ARMv7 SPARC ALPHA	10: MiBench and SPEC-Int 2006	None or Unknown	Cycle
2016	Didehban et al. [9]	gem5	ARMv8	10 MiBench	gem5 Syscall Mode	Instruction
2016	Guan et al. [10] P-FSEFI	QEMU	ARMv7 x86	14 serial NPB	None or Unknown	Instruction
2018	Medeiros et al. [11]	SystemC Model	Plasma-MIPS	26 Malardalen Benchmarks	None or Unknown	Cycle
2022	This work (SOFIA [12])	M*DEV gem5	ARMv6 ARMv7 ARMv8	CIFAR-10 CNN, Mobilenet CNN, 26-Malardalen, 29-NPB, 16-Rodinia	Linux OS, FreeRTOS, and Baremetal	Cycle and instruction

Table 3.2 Summary of fault injection techniques presented in the related works considering virtual platform fault injection simulators. The FI techniques comprise: (A) MMC; (B) Register File; (C) Physical Memory Address; (C) Virtual Memory; (D) Variables and Data Structures; (E) Function Object Code and (F) Function Lifespan

Year	Author/Work	A	B	C	D	E	F	G
2014	Hari et al. [4] Relyzer	✓	✓					
2014	Geissler et al. [5]		✓					
2014	Parasyris et al. [6] GemFI	✓	✓					
2015	Kaliorakis et al. [7] MaFIN and GeFIN	✓	✓					
2016	Tanikella et al. [8] gemV	✓						
2016	Didehban et al. [9]	✓	✓					
2016	Guan et al. [10] P-FSEFI	✓	✓	✓				
2018	Medeiros et al. [11]		✓					
2022	This work (SOFIA [12])		✓	✓	✓	✓	✓	✓

reorder buffer, load-store queue, pipeline queue, renaming unit, and register file. Similarly, Didehban et al. [9] presented a gem5-based fault injection capable of flipping random register file bits, pipeline registers, functional units, and load-store queue. Guan et al. presents the P-FSEFI tool, developed around the QEMU [17] simulator to injects faults in the Central Processing Unit (CPU) logic units, registers, caches, and memory. The experimental setup consists of seven applications from the NAS Parallel Benchmark (NPB) [21], each one in a sequential and a parallel version. Recently, Bandeira et al. [12] presented the SOFIA framework, an extension of the works presented in [14, 22]. The framework's construction considers OVPsim [20] as a base, and the latest version has been enhanced with the Multicore/Multiprocessor Software Development Kit (M*DEV) simulator capabilities [23], which allowed us to incorporate the injection of bit-flips in six different scopes: register file, physical memory, application virtual memory, application variables and data structures, function object code, and function lifespan. Such approaches evaluate soft error reliability considering parallel benchmarks and more comprehensive fault coverage considering not only register files but other micro-architectural elements and software structures. Furthermore, SOFIA flexibility allow us to use the same FI approach and FI flow to evaluate the soft error reliability even in low level RTL[1] processor descriptions.

---

[1] The adopted RTL model must allow to access the register file and memory addressing.

3.1.1   Contribution in Soft Error Assessment Considering Virtual Platforms

Reviewed FI frameworks are also used to investigate different software stacks configurations, such as standard parallelization libraries [24] and compiler optimization flags [11, 25, 26], and their impact on soft error reliability. For instance, authors in [11, 25, 26], investigate the impact of GCC compilation flags (e.g., O0, O3, etc.) on the application behavior under fault influence. These works consider either simple (i.e., in-house and bare-metal applications) or small scenarios, where only a single processor is considered. Modern compilers have specific characteristics, which directly impact on applications performance, power-efficiency, and reliability [27]. Taking a step forward in compiler assessment, Serrano-Cases et al. [28] proposed a method to find out the best combination of compiler optimizations and parameters to improve the fault tolerance of applications.

Aiming to demonstrate the discrepancies between fault injections conducted at different levels, in [29] the authors evaluate the accuracy trade-offs associated with a variety of high-level fault injection techniques (i.e., RTL) comparing to a flip-flop-level baseline. Similarly, the work [30] exploits gate, flip-flop, and ISA level (i.e., register file) FI techniques to evaluate error-rate discrepancies considering the gate-level FI as reference. Cho et al. [29] use geometric means to show the mismatch between the FI levels. Although useful, the authors mention that the adopted metric may not capture how mismatch levels vary across various applications. Aiming at improving the soft error assessment accuracy analysis, authors in [30] use a ranked correlation to evaluate the mismatch between the FI approaches considering the Extrapolated Absolute Failure Count (EAFC) [31] normalized according to the gate-level FI. Such an approach ranks the results from FI techniques of each application, considering the rank shuffling to evaluate the mismatch between the FI approaches. Authors demonstrate that even with discrepancies in such an approach, ISA-level FI approaches are sufficient to evaluate the soft error reliability of low-resource constraint processors (e.g., Cortex-M0). However, the soft error analysis consistency conducted in this work was made based on the EAFC metric. This metric only considers the occurrence of Silent Data Corruptions (SDC)s, which might lead to an inadequate mismatch assessment between different levels of FIs since not all fault classifications are taken into account.

Taking into account the differences between simulators Kaliorakis et al. [7] were the first to be concerned with the accuracy of such FI frameworks based on VPs, comparing FI implementations for gem5 [18] and MARSS [19]. In the same sense, authors in [32] proposed to analyze the soft error rate accuracy of a gem5-based FI framework against results obtained from a neutron beam experiment, considering an Arm Cortex-A9 processor and 13 benchmarks. An initial effort to evaluate the soft error assessment consistency of a JIT-based FI framework is described in [22], where a Fault Injection Module (FIM) was integrated into gem5 and OVPsim. Results considering several fault injection campaigns were compared, and a mismatch of up to 20% is reported. In further experiments, the Authors achieved a lower worst-case mismatch of 12% by reducing the simulation granularity of OVPsim, i.e., the number of instructions executed per simulation cycle. Similar to [7], the conducted experiments aim to evaluate and provide insights on how to improve the accuracy of FI frameworks based on VP simulators.

In this regard, the work conducted in this Book is complementary to [28] that evaluated x86 processors, as it aims to find out which is the best compiler combined with an optimization flag for Arm processors. It also complements [32] because it considers other parameters (e.g., cross-compilers), and the soft error consistency analysis focus on a JIT-based VP. Finally, differs from [22] with regard to the investigation of multi-core systems is that we introduce the discussion on the different cross-compilers.

Even with most reviewed approaches in Tables 3.1 and 3.2 cover several scenarios of fault injection and application benchmark capabilities, *the real consistency of soft error reliability assessment in VP simulators remains open.*

This Book distinguishes from works found in the literature by making an extensive and statistical significance soft error consistency assessment of a JIT-based fault injection framework (the SOFIA OVPsim fault injection module presented in Chap. 5). This work is *the first* to cover so many aspects together, such as: single and multi-core processors (i.e., Arm Cortex-M0, Cortex-M3, and Cortex-A9); two parallel programming models (i.e., Message Passing Interface (MPI) and Open Multi-Processing (OpenMP)) compared with Serial execution; operating system (i.e., FreeRTOS, Linux kernel); five sets of compilers and versions (i.e., GCC 4.9, GCC 7.2, Clang 6, Arm 6.10 and Arm 5.06); optimization flags (i.e., O0, O1, O2, O3, Os and Ofast); and more than 50 applications taken from the NPB [21], Rodinia [33], and the Mälardalen Worst-Case Execution Time (WCET) benchmarks [34]. This range of parameters led to more than 12.7 million fault injections, bringing an excellent confidence level to the results. Note that the soft error consistency evaluation in multi-core processors are presented in [35] since those results were obtained from the experiments conducted at [36].

3.2   Soft Error Reliability Assessment of Machine Learning Techniques

The trend towards having edge computing in our daily lives is to see a wide variety of promising new applications and devices where data collection and analysis are combined [37]. In this regard, lightweight and performance-efficient CNN inference models have been deployed in resource-constrained devices thanks to software libraries and Application Programming Interfaces (APIs) [38]. These libraries/APIs integrate different kernels and quantization techniques [39] that are devoted to reducing traditional machine learning (ML) models' memory and computational requirements.

While the bespoke and optimized implementations of ML models have been studied extensively, the susceptibility of such algorithms to soft errors is still an open research question. In this sense, fault injection campaigns must be conducted to assess the reliability of ML models and understand the fault vulnerability in different approaches [40]. In this direction, recent works conduct fault injections considering software frameworks [41–43], Field-Programmable Gate Arrays (FPGA)s [44–47], Graphic Processing Units (GPU) [48–50], and Application-Specific Integrated Circuits (ASIC) [40, 51].

Aiming to evaluate and improve the reliability of a SVM used in the Mars Odyssey spacecraft, Granat et al. [41] proposed a FI tool built on the Valgrind debugger/profiler [52] called Basic Instrumentation Tool for Fault Localized Injection of Probabilistic SEUs (BITFLIPS). Such a tool was used to validate an Algorithm-Based Fault Tolerance (ABFT) technique implemented in the SVM by injecting faults in the application variables. The experiments demonstrate a significant improvement in system reliability while using ABFT techniques in different SVM kernel functions. The authors in [42] proposed the TensorFI, a FI tool used to evaluate the reliability of TensorFlow ML applications [53]. TensorFI uses a Software Implemented Fault Injection (SWIFI) to inject faults at the inference phase of ML operators in TensorFlow. Furthermore, in [43] the authors proposed the BinFI, an extension of TensorFI, which uses a binary-search technique to pinpoint the safety-critical bits and measure overall resilience in the TensorFlow framework reducing the execution costs. Although both approaches produce significant results, they are restricted to relatively generic software frameworks and do not consider the execution in real devices.

The flexibility of FPGAs enable engineers to design distinct ML models in order to assess different aspects of the soft error reliability in such approaches. For instance, Libano et al. [44] proposed a reliability evaluation methodology for two Feedforward ANNs implemented in an FPGA. Such work started to evaluate the influence of the layer functions complexity in the NN reliability. The experiments consider both radiation-induced and simulation fault injections to evaluate the impact on the *Sigmoid* activation function considering three discretization levels. The study demonstrates that faults occurring at neurons are propagated to multiple instances of the data set having a higher occurrence in hidden layers when compared to output layers. In [45], the authors evaluate the resilience aspects of a typical fully-connected NN accelerator processing the Modified National Institute of Standards and Technology (MNIST) dataset [54]. This case study considers faults randomly injected in specific NN registers while streaming the NN input data. The authors compare the results from fault campaigns with the golden output data and correlate the inference errors in the classification outputs to the application and architecture specifications. Trindade et al. [46] evaluate the soft error reliability of an SVM under both radiation-induced and simulation fault injections experiments. Such an approach assesses the reliability of an FPGA-designed SVM while classifying the fault's criticality. Such study demonstrates that the high number of soft errors in the target SVM architecture critically affects the SVM accuracy. The authors in [47] investigate the soft error effects in a Binarized Neural Network (BNN) accelerator implemented in FPGA considering two topologies: a CNN composed with twelve layers used to classify the Canadian Institute for Advanced Research (CIFAR)-10 dataset [55]; and a Fully-Connected NN with four layers used to classify MNIST dataset [54]. The experiments were performed with a modified version of the Fast Inference for binarized Neural Networks

(FINN) framework [56] targeting weights and activations of the NNs. In this reviewed case, the results demonstrate that the classification accuracy in the BNN accelerator can drastically drop in the presence of soft errors for the worst-case scenarios. Furthermore, the reviewed FPGA-based works started to assess the soft error reliability of ML models considering different scopes, and demonstrate that radiation-induced soft errors might critically affect both the reliability and accuracy of ML inference models.

Reliability becomes an increasing concern as researchers started using CNNs running on GPU Compute Unified Device Architecture (CUDA) cores applied to safety-critical environments. In this sense, Reagen et al. [48] proposed the Ares tool, a fault injection framework that enables the evaluation of soft error reliability of DNNs executing directly on GPUs. Ares is built on top of Keras [57], which takes high-level DNN descriptions using either Theano [58] or TensorFlow [53]. In this work, the fault injections were performed at specific DNN design points, including weights, activations, and hidden states. The authors demonstrate that the soft error reliability varies across the DNN concerning different scopes such as the model, layer type, and data structure. In [49], the authors use SASSI-based Fault Injector (SASSIFI) tool [59] and neutron beam experiments to evaluate the reliability of object detection algorithms in GPUs. The study demonstrates a significant reliability reduction in CNNs once the faults occurring in a GPU tend to propagate to multiple threads. Also, soft errors occurring in CNN layers are most likely to generate critical errors (i.e., errors that could potentially impact safety-critical applications). Similarly, the authors in [50] evaluate layer and kernel vulnerabilities of Residual Network (ResNet) CNN architectures executing on GPUS. This work's experiments show the vulnerability of ResNet kernel and layers while generating a high number of crashes and critical errors. Although GPUs allow a high parallelism and performance capabilities in the execution of CNNs, this advantage generates a high fault propagation through the NN, which means faults are most likely to generate critical errors.

To understand the reliability implications of using high-performance ML accelerators, researchers started to evaluate the fault impact on such devices. In this sense, Li et al. [51] modify the Tiny-DNN framework [60] to inject faults in the buffers of four CNNs running on the specialized Eyeriss DNN accelerator [61]. This work compares the DNN application outputs to evaluate the Failure-in-Time (FIT) rates of the Eyeriss accelerator and classify the output rank deviation (i.e., critical SDCs). The study demonstrates that the sensitivity of the DNN depends on different aspects, such as the network topology and the data type used. Besides that, the buffers implemented to leverage data locality are hugely affected by soft errors, increasing the overall FIT rate of the DNN accelerator. In [40], the authors use radiation-induced experiments to evaluate the reliability of a spontaneously Spiking Neural Network (SNN) from International Business Machines (IBM) TrueNorth neurosynaptic systems [62]. Although the high occurrence of false positives and false negatives in the SNN output classifications, the authors found that the overall classification accuracy remains unaffected.

The impact of SEUs in the execution of ML inference models, particularly convolutional ones has received more attention recently [47, 63, 64]. Such works demonstrated that the soft error reliability of CNN models executing on resource-constrained devices depends on (*i*) the instruction set architecture, (*ii*) the layer where the faults are injected, and (*iii*) the adopted precision bitwidth. Furthermore, other works have also demonstrated that SEUs occurring in data or NN parameters (e.g., weights and activation quantizations) stored in memory affect inference models' soft error reliability and accuracy. Memory faults (e.g., bit flips in memory elements), which may result from environment perturbations and radiation-induced soft errors, may change the stored data (e.g., CNN parameters), which may lead to large deviations of the inference results [48, 51]. In this context, some works aimed at protecting memory cells to ensure the reliability of inference results in the presence of register and memory faults [65–67]. More recently, some works also assess the fault impact considering hardware independent algorithms in TensorFlow backend [68, 69]. In turn, other works investigated the soft error reliability of weights and activations from DNNs [47, 48].

In [51], the authors investigated the soft error effects on datapath registers and buffers of a DNN considering a specialised Tensor Processing Unit (TPU). Reagen et al. [48] started evaluating the soft error reliability of DNN parameters in the training phase through fault injections in bias, weights, activations, and hidden states. The works conducted by [48, 63] assessed the impact of soft errors on weights, bias, and hidden states of DNN models (e.g., CNN) executing on GPUs. In [46], the authors exposed an FPGA board to radiation effects in order to evaluate the soft error reliability in SVM algorithms. Results show that critical faults present high probability to propagate in along the inference. Moreover, the authors in [63] demonstrate that the dataset is critically affected by a single soft error occuring in memory parameters of the CNN executing in GPUs. Similarly, in [47, 70], the analysis comprise the effects of faults on BNN and CNN parameters stored in FPGA based memory. Other works [47, 70, 71] demonstrate that the occurrence of bit-flips in NN's parameters stored in memory affect both reliability and accuracy. Authors have also evaluated the soft error reliability of NN models taking into account different precision implementations varying from floating-point CNNs [63, 70] to 1-bit precision Binarized Neural Networks (BNNs) [47]. Researchers are also investigating new technologies and memory designs aiming at protecting memory cells to reduce disturbances in NN parameters and buffers in the presence of soft errors [65, 67]. The reviewed works demonstrate that a single fault occurring in bias, weights, and activations may cause critical errors compromising the entire classification and causing the majority of the input vectors to be misclassified, which has a serious effect on applications' reliability and accuracy.

The development of ML models targeting the edge is non-trivial due to the limited computational capabilities and energy efficiency of resource-constrained IoT edge devices [72]. In recent years, both academic and industrial researchers have focused their interest on NNs performance and sustainability trade-off, where emerging parallel IoT processors represent an appealing target for tiny ML models since they enable to meet the ML computational constraints [73]. Recent works have been proposed parallel solutions to lift the performance of neural network models [72, 74, 75]. Garofalo et al. [75] provide quantized neural network kernels ranging from 8-bit to 1-bit fixed-point integer inferences targeting a multi-core RISC-V based cluster. In [72], the authors employ non-neural ML kernels to maximize the speedup while using floating-point emulations on a multi-core RISC-V platform. Although underlying approaches enables the execution of ML algorithms on parallel devices, software engineers must consider parallelism as a new issue to be addressed while developing lightweight and performance-efficient software to avoid catastrophes in safety-critical applications. In this sense, while most of the existing works consider HPC or architecture-specific approaches [47–49, 51], the works [49, 76] demonstrate that the parallelization of ML models affects their soft error reliability. Furthermore, the assessment of the soft error reliability at the level of sequential and parallel ML models might be an emergent research question for the next years.

The reviewed works presented different approaches to evaluate several aspects on the soft error reliability of ML models. However, the next step in the soft error reliability assessment is to improve reliability while protect the target platform against radiation effects. Radiation-induced soft errors can be handled either in hardware or software using mitigation techniques. In this sense, the following Sect. 3.2.1 presents a brief review considering system-level soft error mitigation techniques rather than technology-specific approaches that require control of the chip fabrication process, which is often outsourced.

3.2.1 Review of System-Level Soft Error Mitigation Techniques

With the high adoption of nanotechnology manufacturing process the search of methods to reduce the thread of radiation effects in computing systems has become prominent in recent years [77, 78]. Soft error mitigation techniques comprise different methods of protection that might be applied either in hardware, software, or in a hybrid mode. While hardware approaches lead to the area and power overhead, software techniques are generally implemented on a per-application basis that usually incurs performance penalties. Most of such techniques implement any type of redundancy that can vary in terms of time or area/space. The temporal redundancy comprises the same combinational logic used at many separate times, while spatial redundancy replicates the computing element main times [79]. However, soft error mitigation techniques implemented in hardware require physical modification of the target device during the design phase.

Aiming at mitigating soft errors that affects the processor's control-flow and data-flow without changing the chip design, some mitigation techniques are developed on a per-application basis. Such approaches are used due to flexibility and proper fault coverage while charging some performance and energy penalties. For instance, Nicolescu et al. [80] propose an error detection technique that is based on the introduction of data and code redundancy using a set of transformation rules applied to high-level code. In turn, the authors in [81] introduce the REliable Code COmpiler (RECCO), a tool that exploits code reordering and selective variable duplication to generate hardened C/C++ source code automatically. Furthermore the work conducted at [28] use genetic algorithms to find a combination of optimization flags that can increase the final binary reliability while maintaining a reasonable performance and memory utilization trade-off. The authors in [82] developed software-based Triple Modular Redundancy (TMR) and Conditional Modular Redundancy (CMR) mitigation implementations, aiming to reduce the occurrence of soft errors in a Cortex-A9 processor running Linux kernel.

Another software-based alternative to mitigate soft errors comes from low-level code protection. Authors in [83] promote the SoftWare Implemented Fault Tolerance (SWIFT) technique aiming to reduce the overhead associated with Error Detection by Duplicated Instructions (EDDI) [84]. They remove duplicate store instructions, reducing both memory and performance overhead. The SWIFT technique assumes that the system's memory architecture is protected by some error correction mechanism. Results showed a 14% speed-up over EDDI when tested with an Intel Itanium 2. Furthermore, Didehban et al. [9] improved SWIFT technique by checking the load instructions right after a store instruction and creating redundant load instructions in critical sections to achieve near-zero the occurrence of SDC. A popular instruction-level mitigation technique introduced by Reis et al. [85] is the SWIFT-R, which implements a TMR to recover from soft errors in the register file. Instead of duplicating instructions, it triplicates, and change the checking points to a voter mechanism.

In [86], authors presented the Shoestring technique, which exploits a low-cost symptom-based error detection mechanism that focuses on applying instruction duplication to protect only those code segments that are likely to result in user-visible faults and do not exhibit symptomatic behaviour. Results show that Shoestring can recover from an additional 33.9% of soft errors that are undetected by a symptom-only approach. Authors in [87] present the Encore, a software-based error recovery mechanism (paired with other error detection techniques) that combines program analysis, profile data, and simple code transformations to create code portions, which can recover from faults at a minimal cost. Gathered results show that Encore can recover from 97% of transient faults on average with 14% additional runtime overhead. Another TMR-based technique, called ELZAR, is proposed in [88]. It triplicates arithmetic and logical operations, and the voting mechanisms are inserted between register operands of memory and control flow operations for recovery. To reduce the performance overhead introduced by replicated instructions, they utilize Intel Advanced Vector Extensions (AVX), which includes SIMD instructions. The experiments show that the performance overhead is reasonable for CPU-intense applications with many floating-point operations. However, for some case studies, the instruction-level parallelism was inefficient, resulting in a performance penalty that surpassed the SWIFT-R technique.

The NEMESIS technique introduced by Didehban et al. [89] is a duplication with recovery technique. It replicates instructions and checks the results of memory write operations and branches' direction. If an error is detected, it then recovers to a valid state if possible; otherwise, a power restart is needed. The results, obtained from ten selected applications, show that at least 97% of the detected errors are recoverable. Another error recovery technique is the InCheck [90], which is an extension of the near Zero silent Data Corruption (nZDC) technique [9]. The proposed technique comprises of error detection, diagnosis, and recovery schemes. Unlike SWIFT-R, the InCheck mitigates faults by protecting error handling routines in addition to the main program instructions. The authors claim that their technique offers complete error coverage for the tested applications.

3.2.2   Contribution in Machine Learning Soft Error Assessment and Mitigation

Most of the reviewed works consider different approaches to evaluate the reliability of ML models in distinct scopes. Some of the reviewed works evaluate the soft error reliability considering the fault injections in ML operators (e.g., data and parameters) through the early phases of the ML model development framework (i.e., TensorFlow, PyTorch, Keras). Either FPGA and GPU-based works consider both radiation-induced and simulation fault injections to evaluate soft error reliability of different ML models and algorithms. Regarding evaluations, specific points of the analysis stand out considering the layers of the NN and the types of data used in the inference. These characteristics are relevant since they are directly related to the effect of failures both in the reliability and accuracy of ML models. Even with an overall resilience, ASIC-based ML accelerators presented a high occurrence of false positives and false negatives, which can mean catastrophic consequences when applied to safety-critical environments. Table 3.3 shows related works considering soft error assessment in ML models and highlights the approach adopted in this Book. Furthermore, Table 3.4 details the target fault injections and reliability evaluations addressed by the related works.

Emerging IoT systems are expected to incorporate ML models aiming to recognize patterns and predict how systems (e.g., medical) would react to unexpected circumstances [91, 92]. Aiming to enable the execution of such computationally intensive ML models under resource-constrained IoT devices, researchers and industrial leaders are investigating software libraries and APIs. Such libraries/APIs are devoted to support the efficient execution of ML models at reduced memory footprint, which is critical to edge-computing devices [38]. Another approach relies on developing bespoke and optimized ML models, allowing their execution on battery constrained edge IoT devices. The resulting benefit comes at the cost of precision, which has crucial importance in the ML models efficiency and applicability.

In the context of soft error assessment, with the exception of [64] and the work resulting from this Book, reviewed approaches do not consider resource-constraint on their experiments. The majority of these works consider either FPGA implementations of ML models [46, 70, 93] or their execution on GPU [49], DNN accelerators [48, 51, 94] or general-purpose processors [43, 76]. On the soft error mitigation side, traditional partial TMR or

Table 3.3 Related works considering soft error assessment of ML techniques

Author/Year	Environment	FI type	ML model	Datasets
Granat et al. [41]	Framework (Valgrind)	Simulation	SVM	Spacecraft dataset
Li et al. [51]	Eyeriss accelerator	Simulation	AlexNet, CaffeNet, NiN, ConvNet	CIFAR-10, ImageNet
Libano et al. [44]	FPGA	Simulation, radiation-induced	2 Feedforward ANNs	Iris flower, Boston Housing
dos Santos et al. [49]	GPU	Simulation, radiation-induced	YOLO, Faster R-CNN, ResNet	PASCAL VOC 2012, Caltech Pedestrian
Reagen et al. [48]	GPU	Simulation	LeNetFC, LeNetCNN, CF-VGG, VGG16, ResNet50, TiGRU	CIFAR-10, ImageNet, MNIST, TIDIGITS
Salami et al. [45]	FPGA	Simulation	Fully Connected NN	MNIST
Brewer et al. [40]	IBM's TrueNorth	Radiation-induced	SNN	MNIST
Bosio et al. [68]	Framework (Tiny-CNN)	Simulation	LeNet, Tiny YOLO	MNIST, COCO
Rosa et al. [76]	Virtual Platform	Simulation	CNN	KITTI visual odometry
Trindade et al. [46]	FPGA	Simulation, radiation-induced	SVM	150 In-house input vectors
Libano et al. [93]	FPGA	Simulation, radiation-induced	2-layer ANN, 7-layer CNN	Iris flower, MNIST
dos Santos et al. [63]	GPU	Simulation, radiation-induced	YOLO	Caltech pedestrian
Trindade et al. [64]	STM32 L45RE-P board	Radiation-induced	ANN, SVM	Iris flower
Ibrahim et al. [50]	GPU	Simulation	ResNet	ImageNet
Ping et al. [69]	TensorFlow framework	Simulation	LeNet-5, ResNet-50, CIFAR-10 CNN, VGG-16	MNIST, CIFAR-10 CNN, and ImageNet
Khoshavi et al. [47]	FPGA	Simulation	2 BNN: CNN and Fully connected NN	CIFAR-10, MNIST
Luza et al. [70]	FPGA	Radiation-induced	LeNet-5 CNN	MNIST
Chen et al. [95]	GPU	Simulation	LeNet, AlexNet, VGG11, VGG16, ResNet-18, SqueezeNet, Nvidia Dave, Comma.ai	MNIST, Cifar-10, GTSRB, ImageNet, Driving
Adam et al. [96]	GPU	Simulation	AlexNet	ImageNet
Corneliou et al. [71]	FPGA	Emulation	AlexNet	ImageNet
This work	Virtual platform	Simulation	7-layer CIFAR-10 CNN (CMSIS-NN), 29-layer MobileNet (CMix-NN)	CIFAR-10 and ImageNet

specific mitigation techniques have been considered either in FPGA implementations [93] or applied to specialized hardware accelerators [51] or more generic GPUs [49].

Regarding the findings from the reviewed works highlighted in Tables 3.3 and 3.4, soft errors can affect either the reliability and the accuracy of ML models. In this sense, the authors demonstrate that the fault effects vary according to different aspects of the ML model: the model, the topology (i.e., layers and activations), and the dataset (i.e., data type and data optimizations). Different from the above works mentioned in Tables 3.3 and 3.4, this Book *contributes* with the soft error assessment of reduced and mixed precision CNNs

Table 3.4 Summary of addressed FI and reliability evaluation in the related works that aim at assessing the soft error reliability of ML techniques. The target FIs comprise (RF) Register File, (FM) Flash Memory, RAM, (IMA) Isolated Memory Addresses, MMC and (MLD) ML Data. The covered reliability analysis are (OC) Object Code, (TP) Trained Parameters, and (DB) Data Buffers

Year/Work	Target FI						Evaluation			Mitigation	Evaluation
	RF	FM	RAM	IMA	MMC	MLD	OC	TP	DB		
2009 [41]	✓							✓	✓	ABFT	Reliability
2017 [51]	✓	✓	✓						✓		Reliability (Failure-in-Time rates), accuracy
2017 [44]	✓				✓						Reliability (activation functions)
2018 [49]	✓	✓					✓	✓		ABFT	Reliability (safety-critical CNNs on GPUs)
2018 [48]	✓	✓				✓	✓	✓			DNN reliability (model, layers, data structures)
2018 [45]					✓						Reliability, accuracy
2019 [40]	✓	✓									Reliability, accuracy (false positives, false negatives)
2019 [68]	✓					✓		✓			Reliability Memory (weights)
2019 [76]	✓						✓				Correlates multi-core microarchitecture, soft errors
2019 [46]					✓	✓					Reliability, accuracy
2019 [93]							✓	✓		Partial TMR, Full TMR	Reliability (activation functions)
2019 [63]		✓									Reliability of 64 bits, 32 bits, and 16 bits floating point precisions
2020 [64]	✓						✓				Reliability, accuracy
2020 [50]	✓						✓				Reliability
2020 [69]					✓	✓		✓		SEC-DED ECC, Redundant layers	Reliability
2020 [47]		✓						✓			Reliability, accuracy (dataset optimizations)
2020 [70]	✓	✓	✓					✓			Reliability, accuracy (32-bit float, 16-bit and 8-bit integer)
2020 [95]	✓					✓		✓		Selective range restriction	Reliability (Reduce critical SDCs)
2021 [96]	✓				✓			✓			Reliability (Program vulnerability factor)
2021 [71]	✓	✓	✓								Reliability (Average vulnerability factor)
This work	✓	✓	✓	✓			✓	✓	✓	P-TMR and RAT	Reliability and accuracy of mixed low precision ML algorithms targeting resource constrained IoT edge devices

developed with specialized NN kernels (CMSIS-NN and CMix-NN), which are devoted to improve the performance and minimize the memory footprint of NN -based applications. To better understand how soft errors affect the reliability of the ML models, gathered results had been obtained through fault injection campaigns conducted with SOFIA framework [12]. The proposed evaluation comprises both the reduced precision CMSIS-NN [38] kernels, which support 8-bit fixed point quantizations, and CMix-NN [39] kernels, which supports 8, 4, and 2 bit mixed precision quantizations. The adopted case studies are: *(i)* the CIFAR-10 CNN composed of 7 layers developed with CMSIS-NN and trained with CIFAR-10 dataset [55]; and *(ii)* the MobileNet CNN composed of 29 layers developed with CMix-NN and trained with ImageNet dataset [97]. To evaluate the soft error reliability considering the application inference flow, this work employs a fault injection technique that isolates the critical functions of the CNN model's layers. In order to conduct a more relevant assessment, this Book also promotes a fault classification, which considers the impact of critical faults on the output predictions of evaluated ML models. Furthermore, to asses the impact of soft errors in the ML model's parameters (e.g., activation values) storage in memory elements, SOFIA was extended to enable the fault injection in isolated memory sections (i.e., Flash and RAM).

In summary, this Book innovates from previous work in five key directions:

- First, this is the first work to investigate the relationship between the soft error suscep-tibility and reduced precision CNN models, which is completely ignored in the works presented in Tables 3.3 and 3.4;
- Second, this work evaluates the soft error reliability of reduced and mixed precision CNN applications executing on resource-constrained IoT devices, considering different aspects: register file, layers, activations, multithreads, and memory sections;
- Third, this Theses focuses on reducing the occurrence of soft errors in resource-constraint devices. Therefore, this is the first work to evaluate the benefits of using a lightweight technique, RAT, w.r.t. a partial replication technique;
- Fourth, this work explores the relative performance, memory utilization, and soft error reliability trade-offs of two system-level mitigation techniques considering a micropro-cessor running different precision bitwidth variations of MobileNet on ImageNet;
- Fifth, extensive and consistent soft error assessment of adopted case studies considering more than 14.8 million fault injections while maintaining acceptable consistency metrics defined in the literature [98].

References

1. Avižienis, A., Laprie, J.C., Randell, B.: Dependability and its threats: a taxonomy. In: Building the Information Society, pp. 91–120 (2004). https://doi.org/10.1007/978-1-4020-8157-6_13
2. Mansour, W., Velazco, R.: SEU fault-injection in VHDL-based processors: a case study. J. Elec-tron. Test. **29**(1), 87–94 (2013). https://doi.org/10.1007/s10836-013-5351-6

3. Abbasitabar, H., Zarandi, H.R., Salamat, R.: Susceptibility analysis of LEON3 embedded proces-sor against multiple event transients and upsets. In: International Conference on Computational Science and Engineering (CSE), pp. 548–553 (2012). https://doi.org/10.1109/ICCSE.2012.81

4. Hari, S.K.S., Venkatagiri, R., Adve, S.V., Naeimi, H.: Ganges: gang error simulation for hard-ware resiliency evaluation. In: 2014 ACM/IEEE 41st International Symposium on Computer Architecture (ISCA), pp. 61–72 (2014). https://doi.org/10.1145/2678373.2665685

5. de Aguiar Geissler, F., Kastensmidt, F.L., Souza, J.E.P.: Soft error injection methodology based on QEMU software platform. In: 2014 15th Latin American Test Workshop—LATW, pp. 1–5 (2014). https://doi.org/10.1109/LATW.2014.6841910

6. Parasyris, K., Tziantzoulis, G., Antonopoulos, C.D., Bellas, N.: GemFI: A fault injection tool for studying the behavior of applications on unreliable substrates. In: International Conference on Dependable Systems and Networks (DSN), pp. 622–629 (2014). https://doi.org/10.1109/DSN. 2014.96

7. Kaliorakis, M., Tselonis, S., Chatzidimitriou, A., Foutris, N., Gizopoulos, D.: Differential fault injection on microarchitectural simulators. In: International Symposium on Workload Charac-terization (IISWC), pp. 172–182 (2015). https://doi.org/10.1109/IISWC.2015.28

8. Tanikella, K., Koy, Y., Jeyapaul, R., Lee, K., Shrivastava, A.: GemV: a validated toolset for the early exploration of system reliability. In: 2016 IEEE 27th International Conference on Application-Specific Systems, Architectures and Processors (ASAP), pp. 159–163 (2016). https://doi.org/10.1109/ASAP.2016.7760786

9. Didehban, M., Shrivastava, A.: nZDC: A compiler technique for near zero silent data corruption. In: Proceedings of the 53rd Annual Design Automation Conference (DAC), DAC '16, pp. 48:1–48:6. ACM (2016). https://doi.org/10.1145/2897937.2898054

10. Guan, Q., BeBardeleben, N., Wu, P., Eidenbenz, S., Blanchard, S., Monroe, L., Baseman, E., Tan, L.: Design, use and evaluation of P-FSEFI: a parallel soft error fault injection framework for emulating soft errors in parallel applications. In: Proceedings of the 9th EAI International Conference on Simulation Tools and Techniques, SIMUTOOLS'16, pp. 9–17. ICST (Institute for Computer Sciences, Social-Informatics and Telecommunications Engineering) (2016). https:// doi.org/10.5555/3021426.3021429

11. Medeiros, G., Bortolon, F., Ost, L., Reis, R.: Evaluation of compiler optimization flags effects on soft error resiliency. In: Symposium on Integrated Circuits and Systems Design (SBCCI), pp. 1–6 (2018). https://doi.org/10.1109/SBCCI.2018.8533246

12. Bandeira, V., Rosa, F., Reis, R., Ost, L.: Non-intrusive fault injection techniques for efficient soft error vulnerability analysis. In: 2019 IFIP/IEEE 27th International Conference on Very Large Scale Integration (VLSI-SoC), pp. 123–128 (2019). https://doi.org/10.1109/VLSI-SoC.2019. 892037

13. Hari, S.K.S., Adve, S.V., Naeimi, H., Ramachandran, P.: Relyzer: Exploiting application-level fault equivalence to analyze application resiliency to transient faults. In: Proceedings of the Seventeenth International Conference on Architectural Support for Programming Languages and Operating Systems, pp. 123–134. Association for Computing Machinery (2012). https://doi. org/10.1145/2150976.2150990

14. Rosa, F., Kastensmidt, F.L., Reis, R., Ost, L.: A fast and scalable fault injection framework to evaluate multi/many-core soft error reliability. In: International Symposium on Defect and Fault Tolerance in VLSI and Nanotechnology Systems (DFTS), pp. 211–214 (2015). https://doi.org/ 10.1109/DFT.2015.7315164

15. Magnusson, P.S., Christensson, M., Eskilson, J., Forsgren, D., Ha allberg, G., Hogberg, J., Lars-son, F., Moestedt, A., Werner, B.: Simics: A full system simulation platform. Computer **35**(2), 50–58 (2002). https://doi.org/10.1109/2.982916

16. Martin, M.M.K., Sorin, D.J., Beckmann, B.M., Marty, M.R., Xu, M., Alameldeen, A.R., Moore, K.E., Hill, M.D., Wood, D.A.: Multifacet's general execution-driven multiprocessor simulator (GEMS) toolset. SIGARCH Comput. Archit. News **33**(4), 92–99 (2005). https://doi.org/10.1145/1105734.1105747

17. Bellard, F.: QEMU, a fast and portable dynamic translator. In: Proceedings of the Annual Conference on USENIX Annual Technical Conference, vol. 41, issue 46, pp. 10–5555 (2005). https://doi.org/10.5555/1247360.1247401

18. Binkert, N., Beckmann, B., Black, G., Reinhardt, S.K., Saidi, A., Basu, A., Hestness, J., Hower, D.R., Krishna, T., Sardashti, S., Sen, R., Sewell, K., Shoaib, M., Vaish, N., Hill, M.D., Wood, D.A.: The gem5 simulator. SIGARCH Comput. Arch. News **39**(2), 1–7 (2011). https://doi.org/10.1145/2024716.2024718

19. Patel, A., Afram, F., Chen, S., Ghose, K.: MARSS: A full system simulator for multicore x86 CPUs. In: Design Automation Conference (DAC), pp. 1050–1055 (2011). https://doi.org/10.1145/2024724.2024954

20. Imperas: Open Virtual Platforms (OVP) (2021). http://www.ovpworld.org/

21. Bailey, D., Barszcz, E., Barton, J., Browning, D., Carter, R., Dagum, L., Fatoohi, R., Frederickson, P., Lasinski, T., Schreiber, R., Simon, H., Venkatakrishnan, V., Weeratunga, S.: The NAS parallel benchmarks summary and preliminary results. In: Conference on Supercomputing (SC), pp. 158–165 (1991). https://doi.org/10.1145/125826.125925

22. Rosa, F., Ost, L., Reis, R., Davidmann, S., Lapides, L.: Evaluation of multicore systems soft error reliability using virtual platforms. In: International New Circuits and Systems Conference (NEWCAS), pp. 85–88 (2017). https://doi.org/10.1109/NEWCAS.2017.8010111

23. Imperas: DEV—Virtual Platform Development and Simulation (2021). https://www.imperas.com/dev-virtual-platform-development-and-simulation/

24. da Rosa, F., Bandeira, V., Reis, R., Ost, L.: Extensive evaluation of programming models and ISAs impact on multicore soft error reliability. In: Design Automation Conference (DAC), pp. 1–6 (2018). https://doi.org/10.1145/3195970.3196050

25. Lins, F.M., Tambara, L.A., Kastensmidt, F.L., Rech, P.: Register file criticality and compiler optimization effects on embedded microprocessor reliability. IEEE Trans. Nucl. Sci. **64**(8), 2179–2187 (2017). https://doi.org/10.1109/TNS.2017.2705150

26. Sangchoolie, B., Ayatolahi, F., Johansson, R., Karlsson, J.: A study of the impact of bit-flip errors on programs compiled with different optimization levels. In: European Dependable Computing Conference (EDCC), pp. 146–157 (2014). https://doi.org/10.1109/EDCC.2014.30

27. Hoste, K., Eeckhout, L.: COLE: compiler optimization level exploration. In: International Symposium on Code Generation and Optimization (CGO), pp. 165–174 (2008). https://doi.org/10.1145/1356058.1356080

28. Serrano-Cases, A., Morilla, Y., Martín-Holgado, P., Cuenca-Asensi, S., Martínez-Álvarez, A.: Non-intrusive automatic compiler-guided reliability improvement of embedded applications under proton irradiation. IEEE Trans. Nucl. Sci. **66**(7), 1500–1509 (2019). https://doi.org/10.1109/TNS.2019.2912323

29. Cho, H., Mirkhani, S., Cher, C.Y., Abraham, J.A., Mitra, S.: Quantitative evaluation of soft error injection techniques for robust system design. In: Design Automation Conference (DAC), pp. 1–10 (2013). https://doi.org/10.1145/2463209.2488859

30. Schirmeier, H., Breddemann, M.: Quantitative cross-layer evaluation of transient-fault injection techniques for algorithm comparison. In: European Dependable Computing Conference (EDCC), pp. 15–22. Naples, Italy (2019). https://doi.org/10.1109/EDCC.2019.00016

31. Schirmeier, H., Borchert, C., Spinczyk, O.: Avoiding pitfalls in fault-injection based comparison of program susceptibility to soft errors. In: 2015 45th Annual IEEE/IFIP International Conference

on Dependable Systems and Networks, pp. 319–330. Rio de Janeiro, RJ, Brazil (2015). https://doi.org/10.1109/DSN.2015.44

32. Chatzidimitriou, A., Bodmann, P., Papadimitriou, G., Gizopoulos, D., Rech, P.: Demystifying soft error assessment strategies on ARM CPUs: microarchitectural fault injection versus neutron beam experiments. In: International Conference on Dependable Systems and Networks (DSN), pp. 26–38 (2019). https://doi.org/10.1109/DSN.2019.00018

33. Che, S., Boyer, M., Meng, J., Tarjan, D., Sheaffer, J.W., Lee, S.H., Skadron, K.: Rodinia: A benchmark suite for heterogeneous computing. In: International Symposium on Workload Characterization (IISWC), pp. 44–54 (2009). https://doi.org/10.1109/IISWC.2009.5306797

34. Gustafsson, J., Betts, A., Ermedahl, A., Lisper, B.: The mälardalen WCET benchmarks: past, present and future. In: International Workshop on Worst-Case Execution Time Analysis (WCET), pp. 136–146 (2010). https://doi.org/10.4230/OASIcs.WCET.2010.136

35. Abich, G., Garibotti, R., Bandeira, V., da Rosa, F., Gava, J., Bortolon, F., Medeiros, G., Moraes, F.G., Reis, R., Ost, L.: Evaluation of the soft error assessment consistency of a JIT-based virtual platform simulator. IET Comput. Digit. Tech. **15**(2), 125–142 (2021). https://doi.org/10.1049/cdt2.12017

36. da Rosa, F.R.: Early evaluation of multicore systems soft error reliability using virtual platforms. Ph.D. thesis, PGMICRO—UFRGS (2018). https://lume.ufrgs.br/handle/10183/181996

37. Hao, C., Dotzel, J., Xiong, J., Benini, L., Zhang, Z., Chen, D.: Enabling design methodologies and future trends for edge AI: specialization and codesign. IEEE Des. Test **38**(4), 7–26 (2021). https://doi.org/10.1109/MDAT.2021.3069952

38. Lai, L., Suda, N., Chandra, V.: CMSIS-NN: Efficient neural network kernels for arm Cortex-M CPUs (2018). arXiv:1801.06601. https://doi.org/10.48550/arXiv.1801.06601

39. Capotondi, A., Rusci, M., Fariselli, M., Benini, L.: CMix-NN: mixed low-precision CNN library for memory-constrained edge devices. IEEE Trans. Circuits Syst. II: Express Briefs **67**(5), 871–875 (2020). https://doi.org/10.1109/TCSII.2020.2983648

40. Brewer, R.M., Moran, S.L., Cox, J., Sierawski, B.D., McCurdy, M.W., Zhang, E.X., Iyer, S.S., Schrimpf, R.D., Alles, M.L., Reed, R.A.: The impact of proton-induced single events on image classification in a neuromorphic computing architecture. IEEE Trans. Nucl. Sci. **67**(1), 108–115 (2019). https://doi.org/10.1109/TNS.2019.2957477

41. Granat, R., Wagstaff, K.L., Bornstein, B., Tang, B., Turmon, M.: Simulating and detecting radiation-induced errors for onboard machine learning. In: 2009 Third IEEE International Conference on Space Mission Challenges for Information Technology, pp. 125–131 (2009). https://doi.org/10.1109/SMC-IT.2009.22

42. Li, G., Pattabiraman, K., DeBardeleben, N.: Tensorfi: A configurable fault injector for tensorflow applications. In: 2018 IEEE International Symposium on Software Reliability Engineering Workshops (ISSREW), pp. 313–320 (2018). https://doi.org/10.1109/ISSREW.2018.00024

43. Chen, Z., Li, G., Pattabiraman, K., DeBardeleben, N.: Binfi: An efficient fault injector for safety-critical machine learning systems. In: Proceedings of the International Conference for High Performance Computing, Networking, Storage and Analysis (SC), SC '19, pp. 1–23. Association for Computing Machinery, New York, NY, USA (2019). https://doi.org/10.1145/3295500.3356177

44. Libano, F., Rech, P., Tambara, L., Tonfat, J., Kastensmidt, F.: On the reliability of linear regression and pattern recognition feedforward artificial neural networks in FPGAs. IEEE Trans. Nucl. Sci. **65**(1), 288–295 (2017). https://doi.org/10.1109/TNS.2017.2784367

45. Salami, B., Unsal, O.S., Kestelman, A.C.: On the resilience of RTL NN accelerators: fault characterization and mitigation. In: 2018 30th International Symposium on Computer Architecture and High Performance Computing (SBAC-PAD), pp. 322–329 (2018). https://doi.org/10.1109/CAHPC.2018.8645906

46. Trindade, M.G., Coelho, A., Valadares, C., Viera, R.A., Rey, S., Cheymol, B., Baylac, M., Velazco, R., Bastos, R.P.: Assessment of a hardware-implemented machine learning technique under neutron irradiation. IEEE Trans. Nucl. Sci. **66**(7), 1441–1448 (2019). https://doi.org/10.1109/TNS.2019.2920747

47. Khoshavi, N., Broyles, C., Bi, Y.: A survey on impact of transient faults on BNN inference accelerators (2020). arXiv:2004.05915. https://doi.org/10.48550/arXiv.2004.05915

48. Reagen, B., Gupta, U., Pentecost, L., Whatmough, P., Lee, S.K., Mulholland, N., Brooks, D., Wei, G.Y.: Ares: a framework for quantifying the resilience of deep neural networks. In: 2018 55th ACM/ESDA/IEEE Design Automation Conference (DAC), pp. 1–6 (2018). https://doi.org/10.1109/DAC.2018.8465834

49. dos Santos, F.F., Pimenta, P.F., Lunardi, C., Draghetti, L., Carro, L., Kaeli, D., Rech, P.: Analyzing and increasing the reliability of convolutional neural networks on GPUs. IEEE Trans. Reliab. **68**(2), 663–677 (2018). https://doi.org/10.1109/TR.2018.2878387

50. Ibrahim, Y., Wang, H., Bai, M., Liu, Z., Wang, J., Yang, Z., Chen, Z.: Soft error resilience of deep residual networks for object recognition. IEEE Access **8**, 19490–19503 (2020). https://doi.org/10.1109/ACCESS.2020.2968129

51. Li, G., Hari, S.K.S., Sullivan, M., Tsai, T., Pattabiraman, K., Emer, J., Keckler, S.W.: Understanding error propagation in deep learning neural network (DNN) accelerators and applications. In: Proceedings of the International Conference for High Performance Computing, Networking, Storage and Analysis, pp. 1–12 (2017). https://doi.org/10.1145/3126908.3126964

52. Nethercote, N., Seward, J.: Valgrind: a framework for heavyweight dynamic binary instrumentation. ACM Sigplan Not. **42**(6), 89–100 (2007). https://doi.org/10.1145/1273442.1250746

53. Abadi, M., Barham, P., Chen, J., Chen, Z., Davis, A., Dean, J., Devin, M., Ghemawat, S., Irving, G., Isard, M., Kudlur, M., Levenberg, J., Monga, R., Moore, S., Murray, D.G., Steiner, B., Tucker, P., Vasudevan, V., Warden, P., Wicke, M., Yu, Y., Zheng, X.: Tensorflow: a system for large-scale machine learning. In: Proceedings of the 12th USENIX Conference on Operating Systems Design and Implementation, OSDI'16, pp. 265–283. USENIX Association, USA (2016). https://doi.org/10.48550/arXiv.1605.08695

54. Deng, L.: The mnist database of handwritten digit images for machine learning research [best of the web]. IEEE Signal Process. Mag. **29**(6), 141–142 (2012). https://doi.org/10.1109/MSP.2012.2211477

55. Krizhevsky, A., Hinton, G., et al.: CIFAR-10/100—learning multiple layers of features from tiny images. Technical report, University of Toronto (2009). http://citeseerx.ist.psu.edu/viewdoc/download?doi=10.1.1.222.9220&rep=rep1&type=pdf

56. Umuroglu, Y., Fraser, N.J., Gambardella, G., Blott, M., Leong, P., Jahre, M., Vissers, K.: Finn: A framework for fast, scalable binarized neural network inference. In: Proceedings of the 2017 ACM/SIGDA International Symposium on Field-Programmable Gate Arrays, FPGA'17, pp. 65–74. Association for Computing Machinery (2017). https://doi.org/10.1145/3020078.3021744

57. Chollet, F., et al.: Keras: The python deep learning library. Astrophysics source code library pp. ascl–1806 (2018). URL https://github.com/fchollet/keras

58. Al-Rfou, R., Alain, G., Almahairi, A., Angermueller, C., Bahdanau, D., Ballas, N., Bastien, F., Bayer, J., Belikov, A., Belopolsky, A., et al.: Theano: a Python framework for fast computation of mathematical expressions (2016). arXiv:abs/1605.02688. https://doi.org/10.48550/arXiv.1605.02688

59. Hari, S.K.S., Tsai, T., Stephenson, M., Keckler, S.W., Emer, J.: Sassifi: an architecture-level fault injection tool for GPU application resilience evaluation. In: 2017 IEEE International Symposium on Performance Analysis of Systems and Software (ISPASS), pp. 249–258 (2017). https://doi.org/10.1109/ISPASS.2017.7975296

60. Tiny-DNN: Tiny-DNN framework (2017). https://github.com/tiny-dnn/tiny-dnn

61. Chen, Y.H., Emer, J., Sze, V.: Eyeriss: a spatial architecture for energy-efficient dataflow for convolutional neural networks. ACM SIGARCH Comput. Arch. News **44**(3), 367–379 (2016). https://doi.org/10.1145/3007787.3001177

62. Akopyan, F., Sawada, J., Cassidy, A., Alvarez-Icaza, R., Arthur, J., Merolla, P., Imam, N., Nakamura, Y., Datta, P., Nam, G., Taba, B., Beakes, M., Brezzo, B., Kuang, J.B., Manohar, R., Risk, W.P., Jackson, B., Modha, D.S.: Truenorth: design and tool flow of a 65 mw 1 million neuron programmable neurosynaptic chip. IEEE Trans. Comput.-Aided Des. Integr. Circuits Syst. **34**(10), 1537–1557 (2015). https://doi.org/10.1109/TCAD.2015.2474396

63. dos Santos, F.F., Navaux, P., Carro, L., Rech, P.: Impact of reduced precision in the reliability of deep neural networks for object detection. In: 2019 IEEE European Test Symposium (ETS), pp. 1–6 (2019). https://doi.org/10.1109/ETS.2019.8791554

64. Trindade, M.G., Bastos, R.P., Garibotti, R., Ost, L., Letiche, M., Beaucour, J.: Assessment of machine learning algorithms for near-sensor computing under radiation soft errors. In: IEEE International Conference on Electronics, Circuits and Systems (ICECS), pp. 1–4 (2020). https://doi.org/10.1109/ICECS49266.2020.9294938

65. Azizimazreah, A., Gu, Y., Gu, X., Chen, L.: Tolerating soft errors in deep learning accelerators with reliable on-chip memory designs. In: 2018 IEEE International Conference on Networking, Architecture and Storage (NAS), pp. 1–10 (2018). https://doi.org/10.1109/NAS.2018.8515692

66. Guan, H., Ning, L., Lin, Z., Shen, X., Zhou, H., Lim, S.H.: In-place zero-space memory protection for cnn. In: Proceedings of the 33rd International Conference on Neural Information Processing Systems. Curran Associates Inc., Red Hook, NY, USA (2019). https://doi.org/10.48550/arXiv.1910.14479

67. Jasemi, M., Hessabi, S., Bagherzadeh, N.: Enhancing reliability of emerging memory technology for machine learning accelerators. IEEE Trans. Emerg. Top. Comput. pp. 1–7 (2020). https://doi.org/10.1109/TETC.2020.2984992

68. Bosio, A., Bernardi, P., Ruospo, A., Sanchez, E.: A reliability analysis of a deep neural network. In: 2019 IEEE Latin American Test Symposium (LATS), pp. 1–6 (2019). https://doi.org/10.1109/LATW.2019.8704548

69. Ping, L., Tan, J., Yan, K.: SERN: modeling and analyzing the soft error reliability of convolutional neural networks. In: Proceedings of the 2020 on Great Lakes Symposium on VLSI, pp. 445–450. Association for Computing Machinery, New York, NY, USA (2020). https://doi.org/10.1145/3386263.3406938

70. Luza, L.M., Söderström, D., Tsiligiannis, G., Puchner, H., Cazzaniga, C., Sanchez, E., Bosio, A., Dilillo, L.: Investigating the impact of radiation-induced soft errors on the reliability of approximate computing systems. In: 2020 IEEE International Symposium on Defect and Fault Tolerance in VLSI and Nanotechnology Systems (DFT), pp. 1–6 (2020). https://doi.org/10.1109/DFT50435.2020.9250865

71. Corneliou, P., Nikolaou, P., Michael, M.K., Theocharides, T.: Fine-grained vulnerability analysis of resource constrained neural inference accelerators. In: 2021 IEEE International Symposium on Defect and Fault Tolerance in VLSI and Nanotechnology Systems (DFT), pp. 1–6. IEEE (2021). https://doi.org/10.1109/DFT52944.2021.9568281

72. Tabanelli, E., Tagliavini, G., Benini, L.: DNN is not all you need: parallelizing non-neural ML algorithms on ultra-low-power IoT processors (2021). https://doi.org/10.48550/arXiv.2107.09448

73. Qi, X., Liu, C.: Enabling deep learning on IoT edge: approaches and evaluation. In: 2018 IEEE/ACM Symposium on Edge Computing (SEC), pp. 367–372 (2018). https://doi.org/10.1109/SEC.2018.00047

74. Zhang, Y., Du, B., Zhang, L., Wu, J.: Parallel DNN inference framework leveraging a compact RISC-V ISA-based multi-core system. In: Proceedings of the 26th ACM SIGKDD International

Conference on Knowledge Discovery & Data Mining, pp. 627–635 (2020). https://doi.org/10.1145/3394486.3403105

75. Garofalo, A., Rusci, M., Conti, F., Rossi, D., Benini, L.: PULP-NN: accelerating quantized neural networks on parallel ultra-low-power RISC-V processors. Philos. Trans. R. Soc. A **378**(2164), 1–21 (2020). https://doi.org/10.1098/rsta.2019.0155

76. da Rosa, F.R., Garibotti, R., Ost, L., Reis, R.: Using machine learning techniques to evaluate multicore soft error reliability. IEEE Trans. Circuits Syst.-I: Regul. Pap. **66**(6), 2151–2164 (2019). https://doi.org/10.1109/TCSI.2019.2906155

77. Kastensmidt, F.L., Carro, L., da Luz Reis, R.A.: Fault-tolerance techniques for SRAM-based FPGAs, vol. 1. Springer, Berlin (2006). https://doi.org/10.1007/978-0-387-31069-5

78. Avirneni, N.D.P., Somani, A.: Low overhead soft error mitigation techniques for high-performance and aggressive designs. IEEE Trans. Comput. **61**(4), 488–501 (2011). https://doi.org/10.1109/TC.2011.31

79. Mavis, D.G., Eaton, P.H.: Soft error rate mitigation techniques for modern microcircuits. In: 2002 IEEE International Reliability Physics Symposium, Proceedings, 40th Annual (Cat. No. 02CH37320), pp. 216–225. IEEE (2002). https://doi.org/10.1109/RELPHY.2002.996639

80. Nicolescu, B., Velazco, R.: Detecting soft errors by a purely software approach: method, tools and experimental results. In: Design, Automation & Test in Europe Conference & Exhibition (DATE), pp. 57–62 (2003). https://doi.org/10.1007/0-306-48709-8_4

81. Benso, A., Chiusano, S., Prinetto, P., Tagliaferri, L.: A c/c++ source-to-source compiler for dependable applications. In: IEEE/IFIP International Conference on Dependable Systems and Networks (DSN), pp. 71–78 (2000). https://doi.org/10.1109/ICDSN.2000.857517

82. Rodrigues, G.S., Kastensmidt, F.L., Reis, R., Rosa, F., Ost, L.: Analyzing the impact of using pthreads versus OpenMP under fault injection in ARM Cortex-A9 dual-core. In: European Conference on Radiation and Its Effects on Components and Systems (RADECS), pp. 1–6 (2016). https://doi.org/10.1109/RADECS.2016.8093180

83. Reis, G.A., Chang, J., Vachharajani, N., Rangan, R., August, D.I.: SWIFT: software implemented fault tolerance. In: International Symposium on Code Generation and Optimization (CGO), pp. 243–254 (2005). https://doi.org/10.1109/CGO.2005.34

84. Oh, N., Shirvani, P.P., McCluskey, E.J.: Error detection by duplicated instructions in super-scalar processors. IEEE Trans. Reliab. **51**(1), 63–75 (2002). https://doi.org/10.1109/24.994913

85. Reis, G.A., Chang, J., August, D.I.: Automatic instruction-level software-only recovery. IEEE Micro **27**(1), 36–47 (2007). https://doi.org/10.1109/DAC.2018.8465834

86. Feng, S., Gupta, S., Ansari, A., Mahlke, S.: Shoestring: probabilistic soft error reliability on the cheap. ACM SIGARCH Comput. Arch. News **38**(1), 385–396 (2010). https://doi.org/10.1145/1735970.1736063

87. Feng, S., Gupta, S., Ansari, A., Mahlke, S.A., August, D.I.: Encore: low-cost, fine-grained transient fault recovery. In: IEEE/ACM International Symposium on Microarchitecture (MICRO), pp. 398–409 (2011). https://doi.org/10.1145/2155620.2155667

88. Kuvaiskii, D., Oleksenko, O., Bhatotia, P., Felber, P., Fetzer, C.: Elzar: triple modular redundancy using intel AVX (practical experience report). In: IEEE/IFIP International Conference on Dependable Systems and Networks (DSN), pp. 646–653 (2016). https://doi.org/10.1109/DSN.2016.65

89. Didehban, M., Shrivastava, A., Lokam, S.R.D.: NEMESIS: a software approach for computing in presence of soft errors. In: IEEE/ACM International Conference on Computer-Aided Design (ICCAD), pp. 297–304 (2017). https://doi.org/10.1109/ICCAD.2017.8203792

90. Didehban, M., Lokam, S.R.D., Shrivastava, A.: InCheck: an in-application recovery scheme for soft errors. In: ACM/IEEE Design Automation Conference (DAC), pp. 1–6 (2017). https://doi.org/10.1145/3061639.3062265

91. Mahdavinejad, M.S., Rezvan, M., Barekatain, M., Adibi, P., Barnaghi, P., Sheth, A.P.: Machine learning for internet of things data analysis: a survey. Digit. Commun. Netw. **4**(3), 161–175 (2018). https://doi.org/10.1016/j.dcan.2017.10.002

92. Amoh, J., Odame, K.M.: An optimized recurrent unit for ultra-low-power keyword spotting. ACM Interact., Mob., Wearable Ubiquitous Technol. **3**(2) (2019). https://doi.org/10.1145/3328907

93. Libano, F., Wilson, B., Anderson, J., Wirthlin, M.J., Cazzaniga, C., Frost, C., Rech, P.: Selective hardening for neural networks in FPGAs. IEEE Trans. Nucl. Sci. **66**(1), 216–222 (2019). https://doi.org/10.1109/TNS.2018.2884460

94. Kundu, S., Basu, K., Sadi, M., Titirsha, T., Song, S., Das, A., Guin, U.: Special session: reliability analysis for ML/AI hardware (2021). arXiv:2103.12166. https://doi.org/10.48550/arXiv.2103.12166

95. Chen, Z., Li, G., Pattabiraman, K.: A low-cost fault corrector for deep neural networks through range restriction. In: 2021 51th Annual IEEE/IFIP International Conference on Dependable Systems and Networks (DSN) (2021). https://doi.org/10.1109/DSN48987.2021.00018

96. Adam, K., Mohd, I.I., Younis, Y.M.: The impact of the soft errors in convolutional neural network on GPUs: Alexnet as case study. Procedia Comput. Sci. **182**, 89–94 (2021). https://doi.org/10.1016/j.procs.2021.02.012

97. Deng, J., Dong, W., Socher, R., Li, L., Kai, L., Fei-Fei, L.: Imagenet: a large-scale hierarchical image database. In: 2009 IEEE Conference on Computer Vision and Pattern Recognition, pp. 248–255. IEEE (2009). https://doi.org/10.1109/CVPR.2009.5206848

98. Leveugle, R., Calvez, A., Maistri, P., Vanhauwaert, P.: Statistical fault injection: quantified error and confidence. In: Design, Automation and Test in Europe Conference (DATE), pp. 502–506 (2009). https://doi.org/10.1109/DATE.2009.5090716

Soft Error Assessment Methodology

4

This Chapter details the adopted fault injection approaches and their fault injection modules, used to evaluate the soft error resilience of single or multi-core systems (Sect. 4.1). In this sense, each fault injection module is detailed to show its usefulness, that is, RTL (Sect. 4.1.1), gem5 (Sect. 4.1.2.1), and Open Virtual Platform Simulator (OVPsim) (Sect. 4.1.2.2). Next, Sect. 4.2 details the adopted fault classification, which is implemented in the three Fault Injection Modules (FIM)s so that the results are automatically classified. Section 4.3 presents the assessment metrics used in this work, which guarantees the consistency and statistical significance of the results and subsequent conclusions of the research. Finally, Sect. 4.3.1 details the adopted software-based soft error mitigation techniques used in this Book.

4.1 Fault Injection Frameworks

This Section first describes two event-driven fault injection frameworks based on Questa Advanced Simulator (Sect. 4.1.1) and gem5 (Sect. 4.1.2.1), the latter is integrated into the SOFIA framework. Such approaches are used in the present work as *references* for the assessment of single and multi-core systems since RTL descriptions represent the synthesizable real processors and gem5 is known as an accurate simulator for complex multi-core systems [1]. The main functionalities of SOFIA are presented in Sect. 4.1.2.2, and then Sect. 4.1.2.2 describes the OVPsim-based fault injection module.

4.1.1 RTL Fault Injection Module

This Section depicts the developed fault injection module for RTL descriptions, similar to that presented in [2, 3]. The main distinction between FIMs is the moment in which the fault is injected. RTL-FIM executes under a discrete event model of computation (e.g., Questa

© The Author(s), under exclusive license to Springer Nature Switzerland AG 2023 63
G. Abich et al., *Early Soft Error Reliability Assessment of Convolutional Neural Networks Executing on Resource-Constrained IoT Edge Devices*, Synthesis Lectures on Engineering, Science, and Technology, https://doi.org/10.1007/978-3-031-18599-1_4

Advanced Simulator), enabling the injection of faults at any or even within half a clock cycle. Therefore, inserted faults may affect the behavior of both the current and the next instructions. Developed RTL fault injection module explores built-in simulator commands and its observability capability to control and monitor the internal signal of a given processor without requiring any changes in its description.

This approach assumes that the fault injection campaigns comprise the five phases illustrated in Fig. 4.1: (*1*) *Platform Setup*; (*2*) *Golden Reference Model*; (*3*) *Fault Injection Setup*; (*4*) *Fault Injection Simulation*; (*5*) *Fault Analysis*. First, the Platform Setup defines the parameters used in the fault campaign, such as the target architecture and the evaluated application. In the second phase, the FIM executes the target system to extract its behavior under ideal circumstances (i.e., no presence of faults). To improve the performance of the RTL fault injection campaign, this step generates checkpoints from simulation time slices. Then, in the Fault Injection Setup phase, the engine defines the fault configuration, which consists of its location (e.g., register, memory address), position (e.g., register bit), and its insertion time. The fault injection configuration relies on a random uniform function, which

Fig. 4.1 Fault injection campaign flow applicable to the three fault injection approaches, i.e., RTL, gem5 and SOFIA

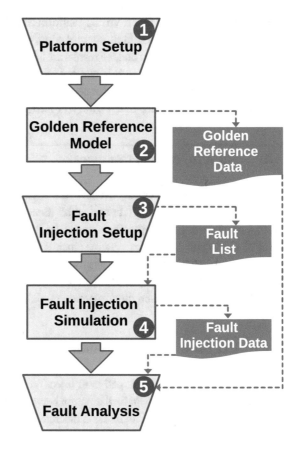

is a well-accepted fault injection technique since it covers the majority of possible faults on a system at a low computation cost [4, 5]. In the Fault Injection Simulation phase, the FIM loads the checkpoint, executes the target system architecture in the presence of the configured faults (i.e., flipped bits), and extracts its behavior. Throughout the Fault Injection Simulation, the developed engine uses available interrupt signals to detect any unexpected activity. Finally, in the last phase, the FIM compares the results against its golden reference data to automatically classify the occurred faults.

4.1.2 SOFIA Framework

Rather than develop a fault injection (FI) framework from scratch, this Book adopted SOFIA—an open-source toolset developed by [6, 7]. The SOFIA framework integrates well-accepted FI techniques along with several facilities (e.g., error tracer module), which enable reliability and software engineers to identify and classify the effects of soft errors on the system's behavior, considering both hardware and software architectures. The adopted framework emulates the occurrence of Single Bit Upsets (SBU)s by injecting faults into pre-selected register or memory locations during the execution of a given software stack (i.e., kernels, drivers, and applications).

In this work, SBUs targets only storage elements due to its higher susceptibility to radiation events when compared to logic elements [8]. Furthermore, register FI modeling is useful to evaluate low-end processors' soft error reliability since it concentrates injection into the more critical structures of the ISA [9]. Nevertheless, the framework supports the injection of bit-flips in the six different scopes presented in Fig. 4.2: (a) register file, (b) physical memory, (c) application virtual memory, (d) application variables and data structures, (e) function object code, and (f) function lifespan. The FI techniques provided in SOFIA can be implemented in any simulation environment that provides access to the system Memory Management Unit (MMU) [6]. Note that in addition to *random register file*, the *function lifespan* technique consider the injection of bit-flips in the register file during the execution of a selected function, while the remaining techniques consider the MMU translation to inject bit-flips in physical memory addresses. The following Sections describe the fault injection modules available in SOFIA framework.

4.1.2.1 SOFIA: Gem5 Fault Injection Module

The SOFIA framework uses the gem5 [10] among the available cycle-accurate virtual platform simulators due to its open and free availability as well as its support for the Arm processor architectures with four CPU models, which differ in speed/accuracy trade-offs. gem5 also supports a rich set of component models, including processor cores, memories, caches, and interconnections. It targets microarchitecture explorations, which incurs sub-

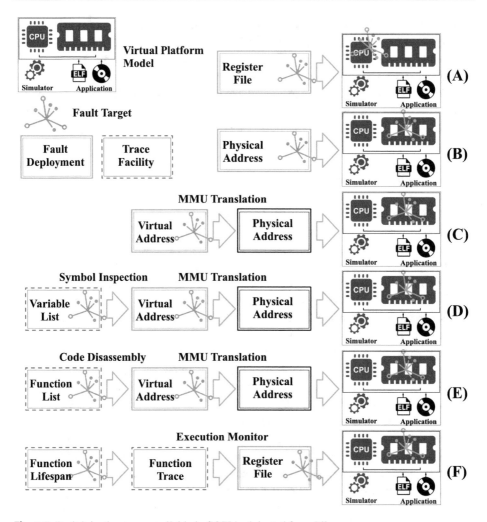

Fig. 4.2 Fault injection types available in SOFIA. Adapted from [6]

stantial simulation overheads due to the number of modeled aspects; typically, it reports simulation performances varying form 2 to 10 Million Instructions Per Second (MIPS).

Additionally, gem5 is a well-known simulator used in many research projects underlying the soft error assessment systems [11–14]. Our gem5-based FIM follows the five-phase fault injection flow illustrated in Fig. 4.1.

Although the phase split is the same for the three FIMs, each one has its own implementation. The first difference w.r.t. RTL is in the platform setup. Here, the application, the kernel, and the configuration of the target architecture are compiled to simulate together in the first phase of the flow. Another difference is on how to inject a fault. While RTL relies on proprietary Questa Advanced Simulator commands such as *force -deposit*, gem5-FIM employs Python scripts to control the simulation flow and uses C/C++ modules to model the microarchitectural components. The deployed fault injection approach minimizes intrusion into the simulator's engines, allowing any researcher in possession of the original simulator to use, modify, or extend its functionality.

4.1.2.2 SOFIA: OVPsim Fault Injection Module

Due to the high simulation speed (typically at hundreds of MIPS), virtual platform simulators based on JIT dynamic binary translation appear to have an advantage over event-driven simulators [15]. The SOFIA framework also considers the M*DEV toolset [16], an advanced version of the OVPsim [17] that includes specific tools to improve the development and verification of embedded software, utilizing virtual platforms. Among available JIT-based virtual platforms, OVPsim [17] distinguishes by its rich number of component models, which includes more than 170 processor variants, memories, Universal Asynchronous Receiver Transmitter (UART), among other components. Note that the JIT-based simulator remains OVPsim and the M*DEV's capabilities are used to enable and improve fault injection techniques and campaigns. In this sense, OVPsim-FIM is an instruction-accurate simulation engine, and thus faults can only occur between instructions. Consequently, injected faults can only influence the behavior of the next instructions. The lack of micro-architecture components and its untimed simulation engine restrict the soft error reliability analysis and impact on the results accuracy. Results demonstrating the loss of accuracy w.r.t. lower level FI simulation approaches is discussed in Sects. 4.1.1 and 4.1.2.1.

OVPsim-FIM also relies on the five-phase fault injection flow shown in Fig. 4.1. However, OVPsim-FIM has an improved fault injection infrastructure, including two simulation techniques to boost the fault injection assessment: checkpoint and an independent parallel simulation engine. The checkpoint technique consists of collecting platform components context during the Golden Reference Model (phase 1) to restore the appropriate context later during the FI campaign, reducing the amount of re-executed code and, consequently, accelerating the simulation time. This technique is built using M*DEV's save and restore functions, allowing restoring the processor and memory context. In addition, the independent parallel simulation technique benefits from the host's processing capacity. The proposed simulation infrastructure comprises a set of C-based functions (e.g., management) and bash scripts developed according to the OVPsim guidelines [16]. The goal is to allocate one platform model per available host-core, enabling the execution of multiple fault injection campaigns in parallel [15]. For example, considering a quad-core processor machine, it is possible to run four platform models injecting 1,000 faults each, reaching 4,000 fault injections in parallel. This being one of the techniques used to enable the assessment of an extremely large number of FI scenarios in this work.

Among the different scopes of bit-flip injections, this work considers the *random register file*, the *function lifespan*, and the *physical memory* to assess complex safety critical applica-

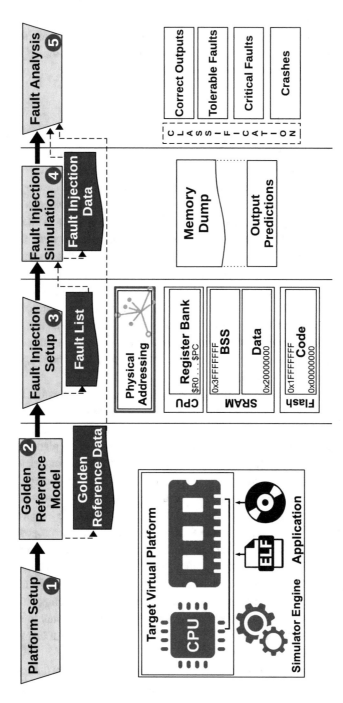

Fig. 4.3 Proposed extension in the physical memory FI technique

tions, for instance, deep inference networks. Aforementioned reviewed works indicate that engineers must be able to early asses the soft error reliability considering both the application execution flow and the storage elements used to accommodate the object code and a given dataset of a neural network application. On the one hand, the *random register file* is a FI technique that homogeneously stimulates all general-purpose registers that execute the CNN application and operating system codes. On the other hand, *function lifespan* reduces the FI spectrum by limiting the insertion time to those small intervals where the target function is active. As the name implies, *physical memory* reproduces faults coming from radiation particles through bit-flip injections into system memory. In the latter technique, to assess the different effects of soft errors occurring in storage elements, this work extends the OVPsim-FIM to inject faults into specific memory sections (e.g., RAM or Flash). Such facility allowed us to assess the reliability of the CNN data, both stored in register files and memory, as later discussed in Sects. 6.1.3 and 6.2.3.

Figure 4.3 illustrates the proposed extension inside the framework FI flow. The developed extension uses the application memory map defined by the compiler to inject faults into the memory section selected in the *Platform Setup* phase. At phase 3, the scripts generate the fault list according to the mapped addressing range. Further, it maps the injection address and the application's output variable to verify and classify the impact of the failure. Note that developed extension is totally non-intrusive, thus not influencing the platform simulation process and consequently without impacting the performance at the fault injection campaign.

4.2 Fault Classification

Aiming to achieve a reasonable confidence level, the three fault injection modules execute the FI campaigns according to an adopted statistical analysis (Sect. 4.3) to characterize the target architecture behavior in the presence of faults, so that additional campaigns do not disturb the result. By default, the gathered results are classified according to the well accepted classification proposed by Cho et al. [5], which defines five possible behaviors for a system in the presence of soft errors:

- *Vanish*: no fault traces are left in both memory and architectural state.
- *Output Not Affected (ONA)*: the resulting memory is not modified, however, one or more remaining bits of the architectural state is incorrect.
- *Output Memory Mismatch (OMM)*: the application terminates without any error indication, and the resulting memory is affected.
- *Unexpected Termination (UT)*: the application terminates abnormally with an error indication.
- *Hang*: the application does not finish requiring a preemptive removal.

Complementary to Cho's classification, this work *proposes* a bespoke fault classification that evaluates the impact of soft errors on the output probabilities of ML case studies. This classification follows the pattern shown in [18–20], where the authors identify the faults in the outputs as: *correct output*, *tolerable faults*, *critical faults*, and *crashes*. As shown in Fig. 4.3, the proposed classification relies on the application output trace during either golden and fault injection simulations. The resulting faulty data is automatically classified according to the following behavior:

– **Correct outputs**: are the scenarios where the output probabilities (e.g., recognized top ranked classification) are the same as faultless execution (i.e., Vanished and ONA);
– **Critical faults**: consider only the OMM faults that affect the output with incorrect probabilities and no predictions (i.e., cases where odds are dispersed, therefore, they have no probabilities in the output data). Incorrect probabilities are those results from OMM, which covers two possibilities: *(i)* when the top-ranked classification differs from the one predicted in the fault-free execution of the ML model; *(ii)* when there is no top-ranked classification (i.e., the odds are dispersed). No prediction cases are those that have no probabilities in the output data.
– **Tolerable faults**: are the remaining OMMs, which are those that have the top-ranked classification equals to the fault-free execution of a NN.
– **Crash**: comprises the application that ends abnormally with an error indication or does not finish, requiring a preemptive removal after a threshold execution time (i.e., effects of UT and Hangs).

Note that such proposed fault classification have been developed based on classification of CNN models. However it can be used for any ML model that saves the output (i.e., probabilities or other either information) after it's execution. Furthermore, it is essential to mention that the main difference between critical fault and crash is that the former refers to a silent error (i.e., the application ends without an error signal) while the latter is a detectable one (i.e., an error signal or unexpected behaviour). Silent errors are considered critical in this work as they can be propagated, which might ultimately incur in human life losses for safety-critical applications (e.g., autonomous vehicles). In contrast, detectable errors can be handled by the system as there is the possibility to reset the system or rerun the algorithm to obtain the correct result.

4.3   System Reliability Assessment Metrics

One of the main concerns when assessing a system's reliability is to develop a precise, well-covered and realistic approach. In this sense, this work sought to ensure that the number of fault injections has a statistical significance by applying the equations developed by [4]. Equation (4.1) shows the minimum number of campaigns needed to cover a certain confidence level with their respective margin of error. While the confidence level assures that if we repeat the experiments, the same results are obtained, and the margin of error indicates the percentage difference between the obtained results and the real value of the population.

$$n = \frac{N}{1 + e^2 \times \frac{N-1}{t^2 \times p \times (1-p)} s} \tag{4.1}$$

where: N is the initial population, p is the estimated probability of a failure (defined as 50%), e is the considered margin of error, and t is the minimum required confidence level (usually 95% is eligible), which is calculated from *t-Distribution* table [21] considering the number of exposed bits as degree of freedom.

Our initial population is the product of possible spatial and temporal fault injections, i.e., the location (e.g., register, memory address) and the position (e.g., register bit) at which the bit-flip will be applied to and its insertion time (e.g., a random clock cycle), respectively. An essential characteristic of the Eq. (4.1) is that when the initial population (N) is large, its increase has little influence on the FI campaign size for a given margin of error and confidence level. For example, for a population greater than 1 million (N), if the chosen confidence level is 99% (in which the calculated t corresponding to 2.575829303549), with a margin of error of 5% ($e = 0.05$), we must perform at least 664 fault injections to provide the necessary confidence in our results.

While maintaining consistency of results, the use of metrics is essential to assess the soft error reliability of a given system. Therefore, this work uses the *Mean Work To Failure (MWTF)* metric [22] to properly assess the soft error reliability impact on the adopted case studies (Chap. 6). Complementary to the fault classification, the MWTF shows the average amount of work that an application can perform until reaching a failure (i.e., higher values are better). This is a fair metric to either compare or evaluate the effects generated by different mitigation techniques. This metric is evaluated in the *Fault Analysis* step (Fig. 4.1). Note that for deep inference networks, the unit work is defined as the relationship between the application's runtime and the most critical vulnerability (i.e., *critical faults*), as shown in Eq. 4.2.

$$MWTF = \frac{1}{(execution\ time \times AVF_{Critical\ Faults})} \tag{4.2}$$

The Architecture Vulnerability Factor (AVF) is used to measure the probability of a fault result in an error (i.e., SDC or Crash) [23]. The AVF critical considers only the SDCs that actually led to wrong classifications. For example, in safety-critical applications, such as autonomous cars, a critical fault can alter the detection of an obstacle in front of the vehicle, which can lead to an accident. For this reason, this work uses the critical-based AVF ($AVF_{critical\ faults}$).

The execution time metric is not ever available in virtual platform simulators such as OVPsim. In this sense, to provide accurate and consistent analysis all case studies have been validated in an of those environments: evaluation board, RTL description, or gem5 simulator. Such environments provide accurate execution time metrics according to the respective platform, which is defined for each case study at the experimental setup (i.e., platform setup phase).

4.3.1   Supported Soft Error Mitigation Techniques

To ensure failsafe functionality of ML-based systems, reliability engineers should be able not only to identify but also explore efficient mitigation solutions to reduce the occurrence of soft errors. SOFIA provides support to 2 mitigation techniques, including *Partial Triple Modular Redundancy (P-TMR)* and *RAT*.

The first mitigation technique is based on a replication approach, i.e., a technique that replicates instructions (except stores and branches) and adds majority voters before conditional branches, load, and store instructions on top of the language-independent Low Level Virtual Machine (LLVM) Intermediate Representation (IR) [24] compilation environment. However, unlike traditional approaches that replicate the entire code, this work uses the *P-TMR* technique that only replicates specific/critical functions, thus minimizing the performance overhead.

The second adopted mitigation technique is the *register allocation technique* (*Register Allocation Technique (RAT)*) [25]. This hardening technique restricts the number of available registers used to execute a specific functions, thus reducing the exposed area. Unlike replication approaches, RAT does not involve code redundancy and is an architecture-independent approach. Also, RAT is a compiler-based mitigating technique; thus, it can be associated with other mitigation techniques.

This work uses the underlying techniques as means to improve the soft error reliability of deep inference models as highly explored in Chap. 6.

References

1. Butko, A., Garibotti, R., Ost, L., Sassatelli, G.: Accuracy evaluation of gem5 simulator system. In: 7th International Workshop on Reconfigurable and Communication-Centric Systems-on-Chip (ReCoSoC), pp. 1–7 (2012). https://doi.org/10.1109/ReCoSoC.2012.6322869
2. Abbasitabar, H., Zarandi, H.R., Salamat, R.: Susceptibility analysis of LEON3 embedded processor against multiple event transients and upsets. In: International Conference on Computational Science and Engineering (CSE), pp. 548–553 (2012). https://doi.org/10.1109/ICCSE.2012.81
3. Bortolon, F.T., Abich, G., Bampi, S., Reis, R., Moraes, F., Ost, L.: Exploring the impact of soft errors on NoC-based multiprocessor systems. In: International Symposium on Circuits and Systems (ISCAS), pp. 1–5 (2018). https://doi.org/10.1109/ISCAS.2018.8351391
4. Leveugle, R., Calvez, A., Maistri, P., Vanhauwaert, P.: Statistical fault injection: quantified error and confidence. In: Design, Automation and Test in Europe Conference (DATE), pp. 502–506 (2009). https://doi.org/10.1109/DATE.2009.5090716

5. Cho, H., Mirkhani, S., Cher, C.Y., Abraham, J.A., Mitra, S.: Quantitative evaluation of soft error injection techniques for robust system design. In: Design Automation Conference (DAC), pp. 1–10 (2013). https://doi.org/10.1145/2463209.2488859

6. Bandeira, V., Rosa, F., Reis, R., Ost, L.: Non-intrusive fault injection techniques for efficient soft error vulnerability analysis. In: 2019 IFIP/IEEE 27th International Conference on Very Large Scale Integration (VLSI-SoC), pp. 123–128 (2019). https://doi.org/10.1109/VLSI-SoC.2019.892037

7. da Rosa, F.R.: Early evaluation of multicore systems soft error reliability using virtual platforms. Ph.D. thesis, PGMICRO—RGS (2018). https://lume.ufrgs.br/handle/10183/181996

8. Seifert, N., Gill, B., Jahinuzzaman, S., Basile, J., Ambrose, V., Shi, Q., Allmon, R., Bramnik, A.: Soft error susceptibilities of 22 nm tri-gate devices. IEEE Trans. Nucl. Sci. **59**(6), 2666–2673 (2012). https://doi.org/10.1109/TNS.2012.2218128

9. Schirmeier, H., Breddemann, M.: Quantitative cross-layer evaluation of transient-fault injection techniques for algorithm comparison. In: European Dependable Computing Conference (EDCC), pp. 15–22. Naples, Italy (2019). https://doi.org/10.1109/EDCC.2019.00016

10. Binkert, N., Beckmann, B., Black, G., Reinhardt, S.K., Saidi, A., Basu, A., Hestness, J., Hower, D.R., Krishna, T., Sardashti, S., Sen, R., Sewell, K., Shoaib, M., Vaish, N., Hill, M.D., Wood, D.A.: The gem5 simulator. SIGARCH Comput. Arch. News **39**(2), 1–7 (2011). https://doi.org/10.1145/2024716.2024718

11. Parasyris, K., Tziantzoulis, G., Antonopoulos, C.D., Bellas, N.: GemFI: a fault injection tool for studying the behavior of applications on unreliable substrates. In: International Conference on Dependable Systems and Networks (DSN), pp. 622–629 (2014). https://doi.org/10.1109/DSN.2014.96

12. Kaliorakis, M., Tselonis, S., Chatzidimitriou, A., Foutris, N., Gizopoulos, D.: Differential fault injection on microarchitectural simulators. In: International Symposium on Workload Characterization (IISWC), pp. 172–182 (2015). https://doi.org/10.1109/IISWC.2015.28

13. Chatzidimitriou, A., Bodmann, P., Papadimitriou, G., Gizopoulos, D., Rech, P.: Demystifying soft error assessment strategies on ARM CPUs: microarchitectural fault injection versus neutron beam experiments. In: International Conference on Dependable Systems and Networks (DSN), pp. 26–38 (2019). https://doi.org/10.1109/DSN.2019.00018

14. da Rosa, F., Bandeira, V., Reis, R., Ost, L.: Extensive evaluation of programming models and ISAs impact on multicore soft error reliability. In: Design Automation Conference (DAC), pp. 1–6 (2018). https://doi.org/10.1145/3195970.3196050

15. Rosa, F., Kastensmidt, F.L., Reis, R., Ost, L.: A fast and scalable fault injection framework to evaluate multi/many-core soft error reliability. In: International Symposium on Defect and Fault Tolerance in VLSI and Nanotechnology Systems (DFTS), pp. 211–214 (2015). https://doi.org/10.1109/DFT.2015.7315164

16. Imperas: DEV—rtual Platform Development and Simulation (2021). https://www.imperas.com/dev-virtual-platform-development-and-simulation/

17. Imperas: Open Virtual Platforms (OVP) (2021). http://www.ovpworld.org/

18. Li, G., Hari, S.K.S., Sullivan, M., Tsai, T., Pattabiraman, K., Emer, J., Keckler, S.W.: Understanding error propagation in deep learning neural network (DNN) accelerators and applications. In: Proceedings of the International Conference for High Performance Computing, Networking, Storage and Analysis, pp. 1 – 12 (2017). https://doi.org/10.1145/3126908.3126964

19. Trindade, M.G., Coelho, A., Valadares, C., Viera, R.A., Rey, S., Cheymol, B., Baylac, M., Velazco, R., Bastos, R.P.: Assessment of a hardware-implemented machine learning technique under neutron irradiation. IEEE Trans. Nucl. Sci. **66**(7), 1441–1448 (2019). https://doi.org/10.1109/TNS.2019.2920747

20. I Khoshavi, N., Broyles, C., Bi, Y.: A survey on impact of transient faults on bnn inference accelerators (2020). arXiv:2004.05915, https://doi.org/10.48550/arXiv.2004.05915
21. McKay, A.: Distribution of the coefficient of variation and the extended "t" distribution. J. R. Stat. Soc. **95**(4), 695–698 (1932). https://www.jstor.org/stable/2342041
22. Reis, G.A., Chang, J., et al.: Software-controlled fault tolerance. ACM Trans. Arch. Code Optim. (TACO) (2005). https://doi.org/10.1145/1113841.1113843
23. Mukherjee, S.S., Weaver, C., Emer, J., Reinhardt, S.K., Austin, T.: A systematic methodology to compute the architectural vulnerability factors for a high-performance microprocessor. In: International Symposium on Microarchitecture (MICRO), pp. 29–40 (2003). https://doi.org/10.1109/MICRO.2003.1253181
24. Lattner, C., Adve, V.: LLVM: a compilation framework for lifelong program analysis transformation. In: International Symposium on Code Generation and Optimization (CGO), pp. 75–86 (2004). https://doi.org/10.1109/CGO.2004.1281665
25. Gava, J., Reis, R., Ost, L.: RAT: A lightweight system-level soft error mitigation technique. In: IFIP/IEEE International Conference on Very Large Scale Integration (VLSI-SOC), pp. 165–170 (2020). https://doi.org/10.1109/VLSI-SOC46417.2020.9344080

Early Soft Error Consistency Assessment

This Chapter presents the evaluation of the soft error assessment consistency considering the SOFIA OVPsim-FIM w.r.t. RTL and gem5 FIMs. This Book contribution lead to the [1] publication. The Sect. 5.1 detail the case studies adopted to assess the consistency of the results considering different software stacks running at single-core resource-constrained architectures. As mentioned before, the soft error results' consistency regarding multi-core architectures are presented in [1] which is separate from this book as they were generated by [2].

5.1 Soft Error Consistency Assessment for Single-Core Processors

This Section aims to thoroughly assess the SOFIA soft error consistency when targeting single-core processors. In this sense, this work considers two real commercial RTL processor descriptions from the Arm Cortex-M family (i.e., Arm Cortex-M0 and Cortex-M3), both available under the Arm University Program [3]. SOFIA natively supports both processor models, and the synbook-ready netlist descriptions of the underlying processors were used to conduct the fault injection campaigns at the RTL level. Section 5.1.1 presents the experimental setup used in the proposed case study. The first results in Sect. 5.1.2 cover the simulation performance brought by the JIT simulator used in SOFIA. Further, in Sect. 5.1.3, we analyze the consistency and the reliability of single-core systems considering different ISAs software stacks (Sect. 5.1.3.1), and cross-compilers (Sect. 5.1.3.2).

5.1.1 Experimental Setup

To provide trustworthy results, conducted experiments consider more than 8.5 million fault injections using 26 applications and varying parameters such as target processor, software

© The Author(s), under exclusive license to Springer Nature Switzerland AG 2023   75
G. Abich et al., *Early Soft Error Reliability Assessment of Convolutional Neural Networks Executing on Resource-Constrained IoT Edge Devices*, Synthesis Lectures on Engineering, Science, and Technology, https://doi.org/10.1007/978-3-031-18599-1_5

Table 5.1 RTL versus SOFIA experimental setup. The * means a sub-flag used in conjunction with other standard flags (e.g., O0, O1, O2, O3)

Processors	Arm Cortex-M0 and Cortex-M3
Software stack	Bare-Metal and FreeRTOS
Benchmark suite	Mälardalen WCET
Number of applications	26
Compilers	GCC 4.9, GCC 7.2, Clang 6, Arm 6.10 and Arm 5.06
Optimization flags	O0, O1, O2, O3, Os, Ofast, Ospace and Otime
Number of compiler sets (with optimization flag)	32
Number of FI campaigns	1140
Injections per campaign	1,000 (Sect. 5.1.3.1) and 10,000 (Sect. 5.1.3.2)
Total fault injections	208,000 (Sect. 5.1.3.1) and 8,320,000 (Sect. 5.1.3.2)

stack, and cross compiler with its optimization flags. Table 5.1 presents the proposed experimental setup used to measure the soft error assessment consistency of SOFIA with respect to the RTL approach.

Selected applications from the Mälardalen WCET benchmark suit [4] include: Adpcm, Binary Search, Bit Manipulation, Blowfish, Bubble Sort, Counts, CRC, Data Compression, Dhrystone, Edn, Exponential Integral, Factorial, Fdct, Fibonacci, Hanoi Tower, Harmonic Calculations, Insert Sort, Jfdctint, Matrix Multiplication, MDC, PeakSpeed, Petri Net, Prime Numbers, Switch Cases, Ud, and Usqrt.

Results were performed on a Linux machine with a Quad-core Intel® Core™i7-7700K CPU and 32 GB DDR4 RAM memory. Fault analyses are obtained by injecting faults (i.e., bit-flips) into the processor's registers (i.e., R0-R15) in a random and uniformly distributed manner, which is widely accepted that circuits are affected [5]. The purpose is to analyze the parameters and Arm processors according to the AVF [5], i.e., a percentage estimate of errors that are not masked, considering the different applications.

5.1.2   FI Simulation Performance of SOFIA w.r.t. RTL

Although electronic hardware engineers usually describe their circuit designs using Hardware Description Languages (HDL) and fault injection at that level seems to be more straightforward due to the accuracy of the results, there are two main reasons for not using HDL models. First, commercial processors are rarely available to Universities and general users in HDL descriptions [6]. Second, the simulation time at this level is exceptionally high. This

Fig. 5.1 Speedup of SOFIA over the RTL-FIM, considering the Cortex-M0 executing 26 benchmarks in bare-metal

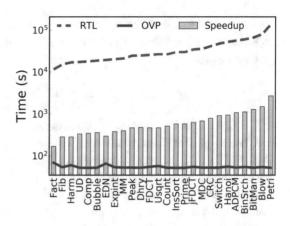

particularity made the simulation speedup of fault injection campaigns one of the leading motivations found in the literature to use virtual platforms [7, 8]. In this sense, our first experiment enable to quantify the simulation speed for each of the adopted applications, as shown in Fig. 5.1.

Figure 5.1 depicts a comparison between the simulation time required to execute a complete fault injection campaign in the RTL and SOFIA approaches. To make straightforward and trustworthy comparisons, the simulation time was extracted considering the Arm Cortex-M0 and bare-metal applications. Results show that SOFIA achieves a remarkable simulation performance, reaching more than three orders of magnitude speedup when compared to the detailed fault injection approach conducted at RTL. Thus, if RTL is considered, thousands of simulations may take several months, which is not suitable to assess the soft error resilience of electronic computing systems. In this context, the utilization of a fault injection JIT-based virtual platform is promising since it can also be used to compare different processor models, ISAs, and benchmarks considering complex Operating Systems (OS) and large scenarios, as shown in Sect. 5.1.3.

5.1.3   Soft Error Reliability Mismatch

This Section presents the accuracy results of the soft error resilience assessment considering single-core processors, comparing SOFIA and RTL-FIM. The following subsections detail each proposed experiment to provide a reasonable and consistency analysis.

5.1.3.1 Mismatch Analysis Considering Different Processor Architectures

The first discussion considers Arm processor architectures, their ISAs and different software stacks to provide us with a solid number of results. In this regard, Eq. (5.1) is used to provide the mismatch between the reference approach (i.e., RTL) and the SOFIA. The mismatch here is defined as the difference between results obtained in RTL and SOFIA, for each application (*app*) and the same fault class (*class*) divided by the number of injected faults in the campaign.

$$Mismatch = \frac{(RTL_{[\ app,\ class\ ]} - SOFIA_{[\ app,\ class\ ]})}{Number\ of\ Injected\ Faults} \tag{5.1}$$

The first results highlight the effects of the Arm Cortex-M0 and Cortex-M3 architectures and their respective ISAs on soft error resilience. Although both processors are from the Cortex-M Family, they differ in terms of computer architectures (i.e., Von Neumann and Harvard, respectively). While the Cortex-M3 has a larger ISA (i.e., entire thumb and thumb-2, 32-bit and 64-bit result multiplication, and 32-bit quotient division), the Cortex-M0 has a reduced ISA (i.e., most thumb, some thumb-2, and 32-bit result multiplication).

Different from [9, 10], our evaluation considers an empirical data distribution that shows the minimum, maximum, and mean mismatch values with the interquartile ranges. Such data distribution enables us to show a detailed inter-benchmark mismatch variation considering all fault classifications and not only the EAFC [10] nor only the overall accuracy [9]. Figure 5.2 shows the mismatch distribution for each fault class, considering the impact of using different processors and software stacks. This means that each bar in Fig. 5.2 represents 26,000 fault injections, which means a confidence level of 99.8% and a error margin of 1%, according to the Eq. (4.1). In addition, the experiments used the *GCC 4.9.3* compiler with *O0* optimization flag, which facilitates the reproducibility of the experiments by other researchers.

First, results show that the architectural difference between the two processors affects the soft error system's reliability, producing different fault class behaviors, as shown in Fig. 5.2. A significant architectural difference is that the Arm Cortex-M0 has a reduced instruction set, which demands the execution of a broader set of instructions to complete more complex operations than the Cortex-M3. For instance, while a 32-bit multiplication operation may vary from 1 up to 32 cycles in the Cortex-M0, the Cortex-M3 multiplier instruction needs a single cycle to complete the same operation.

From a bird's eye view, Fig. 5.2 shows that results obtained from SOFIA differ more prominently, from the reference, in the occurrence of *ONA* and *UT* faults, whereas the mean is out of ±5% (i.e., a low mismatch reference). In this regard, the RTL approach is more likely to either mask an injected fault targeting a general-purpose register (i.e., *Vanished*) or propagate it to other registers (e.g., *ONA* and *UT*). In addition, the results differ due to the simulation nature of each fault injection approach. In SOFIA, faults are injected between instructions, which can restrict their propagation and, consequently, reduce the cumulative effects of soft errors. In turn, the nature of discrete event simulators allows the injection into any clock cycle, which can affect the execution of multi-cycle instructions (e.g., a division execution can take 2–12 cycles on the Cortex-M3). In this case, the value of a register can change a *mid* instruction, which may lead to either a masked fault; or an undefined processor state; or even wrong application execution.

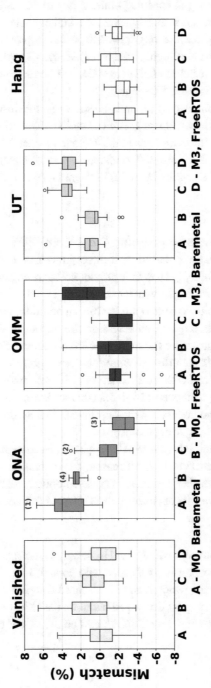

Fig. 5.2 Boxplot of fault mismatch between SOFIA and RTL-FIM considering different architectures and software stacks. Outlier cases (represented by circles) are at least two standard deviations from the mean

On the other hand, looking at specific points, it is possible to infer that the mismatch distribution between bare-metal scenarios revealed a higher occurrence of *ONA* (i.e., mean value indicated by (1) in Fig. 5.2) for the Cortex-M0 on RTL approach, and a lower mismatch with Cortex-M3 (i.e., mean value highlighted by (2) in Fig. 5.2). Also, the results show that SOFIA presents lower mismatches of *Hang*, *OMM*, and *Vanish* faults for both processors, i.e., most mean values are in the $\pm 5\%$ range.

Results in Fig. 5.2 (highlighted by (3) and (4)) show that the mismatch distribution from *ONA* faults is more disperse when using SOFIA, which affects the other fault types proportionally, i.e., the sum of the variations in the average mismatch between bare-metal and FreeRTOS is equal to zero. The collected results show that the mean values remain between $\pm 4\%$, confirming that the inclusion of FreeRTOS did not affect the SOFIA soft error assessment accuracy.

5.1.3.2 Mismatch Analysis Considering Cross-Compilers

Compilers play an essential role due to their direct impact on applications performance, power-efficiency, and reliability [11], as they provide software engineers with a wide variety of optimization settings (i.e., flags), which can be used to either configure debugging and warning messages or to achieve code optimization. In addition, industrial leaders employ different compilers in their projects, so assessing the impact of these compilers on soft error reliability is vital to guarantee the success of their products. On the one hand, most of the work on compilation flags in the literature focuses on performance optimization [12], and on the memory usage and code size reduction [13]. Few are those that assess the soft error reliability provided by compilers [14–16]. In this scenario, this Section investigates the impact of widely adopted compilers and their optimization flags on the soft error reliability to find the most reliable set for Arm processors.

The long simulation time inherent to RTL approach restricts its use for the soft error assessment of multiple scenarios. For that reason, the evaluation considers only the Arm Cortex-M3 to investigate whether the nature of cross-compilers and optimization flags affect the accuracy of soft error results obtained with SOFIA. In this sense, five cross-compilers are considered:

- *GCC 4.9.3*: free-software, still widely used by legacy systems and applications;
- *GCC 7.2.1*: free-software, currently shipped with popular Linux distributions;
- *Clang 6.0.1*: free-software, which is an LLVM-based compiler;
- *Arm 5.06*: proprietary-compiler developed based on the GCC compiler;
- *Arm 6.10*: proprietary-compiler developed based on the LLVM compiler.

Table 5.2 Mean and worst-case absolute mismatch percentages for Arm compiler versions and their optimization flags in Arm Cortex-M3. The * means a combined flag equivalent to *Ofast*

		Arm 5.06								Arm 6.10					
		O0	O0	O1	O1	O2	O2	O3	O3	O0	O1	O2	O3	Os	Ofast
Vanished	Mean	1.5	1.6	1.5	1.6	1.6	1.8	1.6	1.8	1.4	1.7	2.0	1.9	1.8	2.1
	Worst	4.2	6.1	4.7	4.2	5.6	5.1	4.9	5.9	7.5	4.9	5.4	5.9	5.2	5.1
ONA	Mean	1.1	1.1	1.1	1.4	1.5	1.6	1.3	1.6	1.8	2.9	1.6	1.2	1.7	1.3
	Worst	3.6	5.3	4.6	4.1	4.9	4.9	5.0	6.0	6.8	7.1	5.6	6.2	6.0	5.4
OMM	Mean	1.2	1.5	1.1	1.8	1.3	2.0	1.2	1.6	1.2	1.9	1.4	1.3	1.4	1.3
	Worst	4.0	4.5	3.8	5.5	3.1	5.2	3.2	5.4	4.8	5.5	6.4	3.5	3.3	5.1
UT	Mean	1.6	2.6	1.2	2.4	1.2	2.0	1.4	1.4	1.3	1.7	1.5	1.7	1.7	1.9
	Worst	6.1	5.9	5.2	6.6	5.7	5.6	5.2	3.1	3.4	4.3	4.2	3.9	4.5	4.5
Hang	Mean	1.3	1.6	1.4	1.6	1.4	2.2	1.7	1.9	2.0	1.5	1.7	1.5	1.7	1.6
	Worst	4.1	3.8	4.4	4.4	4.6	4.4	5.8	4.7	6.0	3.6	4.4	3.4	4.3	3.5

These compilers consider six optimization flags (i.e., *O0*, *O1*, *O2*, *O3*, *Os*, and *Ofast*), with the exception of the Arm 5.06 compiler, which has a different approach for *Os* and *Ofast*. In this case, the traditional flags *O0*, *O1*, *O2*, and *O3* are combined with other flags (i.e., *Ospace* and *Otime*), which are equivalent to either *Os* and *Ofast*, resulting in eight flag combinations.

Tables 5.2, 5.3, and 5.4 present a mismatch percentage summary of the compiler sets (i.e., compiler with optimization flag) between the RTL and SOFIA FI modules. Considering only bare-metal applications, each compiler set comprises 26,000 fault injection campaigns, which means that our results have a confidence level of 99.8% and a margin of error of 1%, according to Eq. (4.1). The results show that compilers have an impact on soft error resilience. However, the mismatch between the two approaches is very low, with means close to 2 for all compiler sets.

Regarding the mismatch distribution, worst-case scenarios maintain the absolute mismatch below 8%. Considering that the calculated margin of error is at 1%, even the outlier values are very close between the two FI approaches, which means that SOFIA provides good reliability results with orders of magnitude faster than RTL. In turn, comparing the compilers, the results of the two GCCs show that they have the highest number of outliers in worst-cases (see red values in Tables 5.2, 5.3, and 5.4). These outliers may not yield an apparent correlation, but they present a difference in terms of executed instructions. For example, the GCC versions of the Dhrystone application execute 1.7× more instructions than the version generated by the Arm 5.06 compiler, which affects the application vulnerability window.

After presenting an overview of the compilers and optimization flags showing the accuracy of the results in relation to RTL, we increased fault injection campaigns to 10,000 for each compiler set and compared them using SOFIA. The goal is to find the most reliable compiler set, so the FI campaigns were increased to produce a more significant result in terms of confidence level and lower error margin. In this scenario, each compiler set has

Table 5.3 Mean and worst-case absolute mismatch percentages for the Clang compiler version and their optimization flags in Arm Cortex-M3

	Clang 6.0.1						
		O0	O1	O2	O3	Os	Ofast
Vanished	Mean	0.9	1.7	1.5	1.5	1.9	1.6
	Worst	2.6	4.6	4.1	3.9	5.5	3.8
ONA	Mean	1.8	2.0	1.5	1.5	1.7	1.7
	Worst	6.2	5.2	6.3	4.4	3.8	5.9
OMM	Mean	1.2	2.0	2.2	2.0	1.8	2.4
	Worst	7.3	6.9	5.1	4.9	4.7	5.0
UT	Mean	1.9	1.4	2.0	1.6	1.6	1.7
	Worst	4.5	5.3	5.5	5.5	6.5	5.4
Hang	Mean	1.9	1.0	0.8	0.8	1.6	0.7
	Worst	5.4	2.8	2.5	2.4	5.4	2.3

Table 5.4 Mean and worst-case absolute mismatch percentages for the GCC compiler versions and their optimization flags in Arm Cortex-M3

		GCC 4.9.3						GCC 7.2.1					
		O0	O1	O2	O3	Os	Ofast	O0	O1	O2	O3	Os	Ofast
Vanished	Mean	1.8	2.3	2.0	2.1	1.7	1.9	1.4	1.4	1.3	1.8	1.8	1.5
	Worst	4.3	6.7	7.5	6.8	5.6	3.9	3.2	4.4	8.0	6.5	6.8	5.3
ONA	Mean	1.4	2.4	1.8	1.7	1.9	2.2	1.4	1.6	1.1	1.7	1.7	1.8
	Worst	3.6	6.3	5.9	5.8	5.8	6.9	4.5	4.0	6.8	6.2	4.9	5.3
OMM	Mean	2.4	2.2	1.7	1.8	1.9	1.7	2.2	2.0	1.2	1.6	1.9	1.8
	Worst	7.2	6.8	6.5	6.3	5.0	4.0	5.6	6.1	4.5	6.1	6.2	6.0
UT	Mean	3.1	1.7	2.4	2.2	2.6	1.7	2.6	2.3	1.6	1.8	2.1	1.9
	Worst	7.1	6.0	6.3	4.5	7.4	6.3	5.6	5.0	6.0	3.9	4.9	3.9
Hang	Mean	1.5	1.8	1.6	1.6	1.8	2.2	1.6	1.8	1.5	1.6	1.4	1.4
	Worst	3.2	4.8	4.5	4.1	6.9	4.5	5.0	4.7	4.9	4.3	4.8	4.1

260,000 FI campaigns, which leads to a confidence level of 98% and a 0.2% of margin of error.

Figure 5.3 shows the soft errors related to the fault injection campaigns for the five compilers and their optimization flags using SOFIA. This evaluation considers AVF to be the percentage of the sum of all faults that could be analyzed (i.e., *ONA, OMM, UT*, and *Hang*). Thus, the reliability is related to the percentage of *Vanishes* found. Among the evaluated compilers, the commercial *Arm 6.10* presents more occurrences of *Vanish*, indicating that it has a greater soft error resilience. From the open-source alternatives, *Clang* appears to be the best option due to two main reasons: (*i*) a higher number of vanishes were identified, and (*ii*) its adoption leads to a lower occurrence of *Hangs*. On the other hand, the GCC compilers followed the trend shown in Tables 5.2, 5.3, and 5.4, and continued to present the

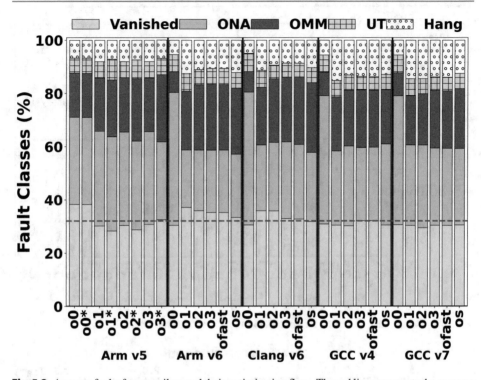

Fig. 5.3 Average faults for compilers and their optimization flags. The red line represents the average number of vanished faults

worst results in terms of soft error resilience, where no optimization flag passed the dashed red line shown in Fig. 5.3.

To facilitate this view, Fig. 5.4 is a zoom-in considering only the occurrence of *Vanishes*. Although the newer GCC version uses more specific Arm Cortex-M3 ISA instructions, the resulting improvement is negligible compared to the previous version. In short, compilers with better results regarding *Vanish* are LLVM-based. Although restricted to only two ISAs, these findings suggest a pattern for adopting LLVM-based compilers, but additional experiments considering other ISAs would be necessary to further explore the reliability benefits of this compiler.

Lastly, we believe that the industry will not adopt a compiler set just because it is more reliable, except for a niche such as critical-safety applications (e.g., autonomous vehicle applications). In this sense, we propose to evaluate which compiler set is more reliable and also offers the best performance. Figure 5.5 shows a trade-off between performance (i.e., execution time from RTL and executed instructions from SOFIA) and reliability (i.e., the occurrence of *Vanishes*), considering the aforementioned cross-compilers. The upper-left corner presents the best of both performance and reliability; conversely, the lower-right

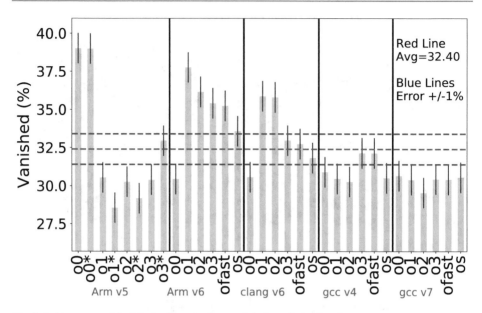

Fig. 5.4 Average vanished faults for compilers and their optimization flags

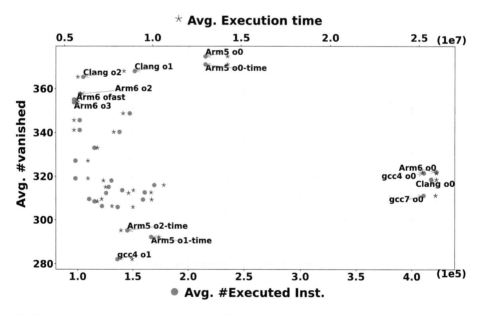

Fig. 5.5 Average execution time and executed instructions versus number of Vanishes considering different compiler sets

corner has the worst performance and reliability. In turn, the lower-left corner assemblies are the configurations that show reasonable performance but low soft error reliability.

Results show that the applications compiled with the *O0* flag presented a higher suscepti- bility to soft errors and a low performance, except for the codes generated with the *Arm 5.06* compiler, which showed a reliability improvement but still below the other sets of compilers. At the other end are the *Arm 6.10* and *Clang 6.0.1* compilers (i.e., both LLVM-based), the two have the best set of performance and reliability. Among them, the *O2* optimization flag stands out, presenting superior results for the two compilers. However, due to the signifi- cance of our results (high confidence level and low margin of error), *Clang* has a certain advantage, and it would be our compiler suggestion. Finally, the remaining compiler sets maintain similar results to the mean of this trade-off.

5.1.4 Closing Remarks

This Chapter demonstrated that the use of virtual platforms brings orders of magnitude speedups to FI campaigns. This also showed that the architectural difference between the two Arm processors affects the system's reliability, producing different fault class behaviors. However, the inclusion of an operating system did not affect the soft error assessment accuracy. Next, after a solid and extensive soft error reliability assessment considering a variety of off-the-shelf compilers, it is possible to conclude that the best compilers are LLVM-based and that the best set in terms of performance and reliability would be the *Clang 6.0.1* compiler using the *O2* optimization flag.

References

1. Abich, G., Garibotti, R., Bandeira, V., da Rosa, F., Gava, J., Bortolon, F., Medeiros, G., Moraes, F.G., Reis, R., Ost, L.: Evaluation of the soft error assessment consistency of a JIT based virtual platform simulator. IET Comput. Digit. Tech. **15**(2), 125–142 (2021). https://doi.org/10.1049/cdt2.12017
2. da Rosa, F.R.: Early evaluation of multicore systems soft error reliability using virtual platforms. Ph.D. thesis, PGMICRO—UFRGS (2018). https://lume.ufrgs.br/handle/10183/181996
3. ARM: Arm University Program (2020). http://www.arm.com/support/university
4. Gustafsson, J., Betts, A., Ermedahl, A., Lisper, B.: The mälardalen WCET benchmarks: past, present and future. In: International Workshop on Worst-Case Execution Time Analysis (WCET), pp. 136–146 (2010). https://doi.org/10.4230/OASIcs.WCET.2010.136
5. Mukherjee, S.S., Weaver, C., Emer, J., Reinhardt, S.K., Austin, T.: A systematic methodology to compute the architectural vulnerability factors for a high-performance microprocessor. In: International Symposium on Microarchitecture (MICRO), pp. 29–40 (2003). https://doi.org/10.1109/MICRO.2003.1253181
6. DeVoe, M.: An industry-university development tools collaboration (2015). https://issuu.com/opensystemsmedia/docs/ecd_november_2015_issuu

7. Kaliorakis, M., Tselonis, S., Chatzidimitriou, A., Foutris, N., Gizopoulos, D.: Differential fault injection on microarchitectural simulators. In: International Symposium on Workload Characterization (IISWC), pp. 172–182 (2015). https://doi.org/10.1109/IISWC.2015.28

8. Rosa, F., Ost, L., Reis, R., Davidmann, S., Lapides, L.: Evaluation of multicore systems soft error reliability using virtual platforms. In: International New Circuits and Systems Conference (NEWCAS), pp. 85–88 (2017). https://doi.org/10.1109/NEWCAS.2017.8010111

9. Cho, H., Mirkhani, S., Cher, C.Y., Abraham, J.A., Mitra, S.: Quantitative evaluation of soft error injection techniques for robust system design. In: Design Automation Conference (DAC), pp. 1–10 (2013). https://doi.org/10.1145/2463209.2488859

10. Schirmeier, H., Breddemann, M.: Quantitative cross-layer evaluation of transient-fault injection techniques for algorithm comparison. In: European Dependable Computing Conference (EDCC), pp. 15–22. Naples, Italy (2019). https://doi.org/10.1109/EDCC.2019.00016

11. Hoste, K., Eeckhout, L.: COLE: Compiler optimization level exploration. In: International Symposium on Code Generation and Optimization (CGO), pp. 165–174 (2008). https://doi.org/10.1145/1356058.1356080

12. Machado, R.S., Almeida, R.B., Jardim, A.D., Pernas, A.M., Yamin, A.C., Cavalheiro, G.G.H.: Comparing performance of C compilers optimizations on different multicore architectures. In: International Symposium on Computer Architecture and High Performance Computing Workshops (SBAC-PADW), pp. 25–30 (2017). https://doi.org/10.1109/SBAC-PADW.2017.13

13. Song, L., Feng, M., Ravi, N., Yang, Y., Chakradhar, S.: COMP: Compiler optimizations for manycore processors. In: International Symposium on Microarchitecture, pp. 659–671 (2014). https://doi.org/10.1109/MICRO.2014.30

14. Lins, F.M., Tambara, L.A., Kastensmidt, F.L., Rech, P.: Register file criticality and compiler optimization effects on embedded microprocessor reliability. IEEE Trans. Nucl. Sci. **64**(8), 2179–2187 (2017). https://doi.org/10.1109/TNS.2017.2705150

15. Medeiros, G., Bortolon, F., Ost, L., Reis, R.: Evaluation of compiler optimization flags effects on soft error resiliency. In: Symposium on Integrated Circuits and Systems Design (SBCCI), pp. 1–6 (2018). https://doi.org/10.1109/SBCCI.2018.8533246

16. Serrano-Cases, A., Morilla, Y., Martín-Holgado, P., Cuenca-Asensi, S., Martínez-Álvarez, A.: Non-intrusive automatic compiler-guided reliability improvement of embedded applications under proton irradiation. IEEE Trans. Nucl. Sci. **66**(7), 1500–1509 (2019). https://doi.org/10.1109/TNS.2019.2912323

Soft Error Reliability Assessment of ML Inference Models Executing on Resource-Constrained IoT Edge Devices

6

This Chapter provides in-depth insights and results related to the second main contribution of this Book: the early soft error assessment and mitigation of ML inference models executing on resource-constrained IoT edge devices. The underlying contribution has been published in several international conferences [1–3] and high quality journals [4, 5]. In this sense, different soft error reliability assessments of ML models build from industrial libraries and APIs targeting resource-constrained IoT edge systems are detailed discussed. The discussion is based on reasonable number of results obtained from a vast number of FI campaigns, taking into account different reliability aspects of a NN model, such as topology, layers, target ISA optimizations, memory parameters, precision bitwidth, mitigation techniques, and parallelization.

Table 6.1 details the target evaluation according to each executed CNN model. Section 6.1 presents the early soft error assessment considering the CIFAR-10 CNN execution in resource-constrained IoT edge devices. Furthermore, Sect. 6.2 comprise the soft error reliability results considering the MobileNet CNN with different bitwidth configurations.

6.1 Soft Error Reliability Assessment of the CIFAR-10 CNN

This section presents the soft error reliability assessment results for the CIFAR-10 CNN case study. First, Sect. 6.1.1 presents a detailed description of the CIFAR-10 CNN developed with the CMSIS-NN library. Next, Sects. 6.1.2, 6.1.3, and 6.1.4 comprise the soft error reliability assessment considering the CIFAR-10 CNN execution, topology, isolated layers, target ISA optimizations, memory (code, parameters and data), and thread parallelism.

© The Author(s), under exclusive license to Springer Nature Switzerland AG 2023
G. Abich et al., *Early Soft Error Reliability Assessment of Convolutional Neural Networks Executing on Resource-Constrained IoT Edge Devices*, Synthesis Lectures on Engineering, Science, and Technology, https://doi.org/10.1007/978-3-031-18599-1_6

Table 6.1 Contributions organized by case study

	ML model	
Target evaluation	CIFAR-10 CNN	MobileNet CNN
Topology	✓	✓
Layers	✓	✓
ISA	✓	
Memory	✓	✓
Precision bitwidth		✓
Soft error mitigation		✓
Parallelism	✓	

6.1.1 CIFAR-10 CNN Developed with CMSIS-NN

This Section presents the CIFAR-10 case-study, which consists of a 7-layer CNN developed with the CMSIS-NN kernels [6] to run in Common Microcontroller Software Interface Standard (CMSIS) supported processors (e.g., Arm Cortex-M). The CNN is trained with the CIFAR-10 dataset [7], consisting of 60,000 32 × 32 color images divided into ten output classes. Figure 6.1 shows an overview of a typical CNN topology based on the built-in example provided in CMSIS-NN [6]. Table 6.2 details the CNN layers with the kernel shapes (i.e., filter and output) used in each layer. Such CNN consists of multiple layers composed of convolution layers, interspersed by non-linear activation layers, pooling layers, and a fully-connected layer at the end of CNN.

The overview of CMSIS-NN kernel is shown in Fig. 6.2. The kernel code consists of two parts: *NNFunctions* and *NNSupportFunctions*. As described in [6], the NNFunctions includes the functions that implement popular neural network layer types while the NNSupportFunctions includes the utility functions.

Different from the traditional NN models trained using 32-bit floating-point data representation, the CMSIS-NN uses low-precision fixed-point representation as high precision is generally not required during the inference. The use of such quantization avoids the need for floating-point de-quantization between layers, as some Arm Cortex-M processors do not have a dedicated Floating-Point Unit (FPU). The CMSIS-NN kernels deploy optimizations that aim to boost performance while reducing the memory-footprint when executing complex NN inferences in resource-constrained processors. Such optimizations focus on enabling neural networks on Cortex-M based systems that support SIMD instructions, especially 16-bit Multiply-And-Accumulate (MAC) instructions, such as Signed Multiply Accumulate Long Dual (SMLAD), which are very useful for NN computation. In this sense, the CMSIS-NN kernel implements matrix multiplications with a 2 × 2 dimension to enable some data reuse and reduce the number of load instructions. Further, it performs accu-

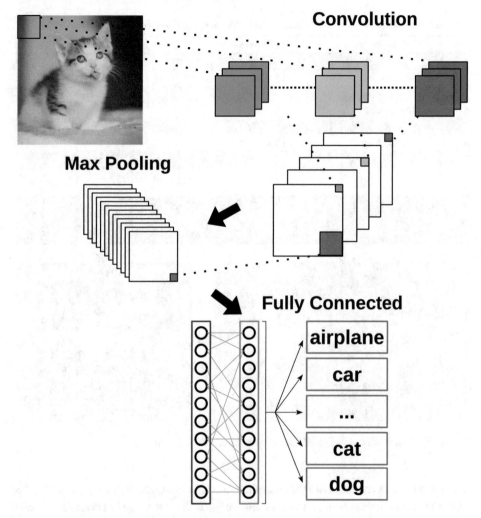

Fig. 6.1 Brief illustration of the adopted CIFAR-10 CNN topology. Adapted from [6]

mulation with dedicated SIMD MAC instruction, and data transformation occurs without reordering, achieving reasonable performance [6].

To achieve such a performance, the convolution layer implements a partial image-to-column data transformation considering the Height-Width-Channel (HWC) format. Rectified Linear activation Unit (ReLU) layer implements a loop over all elements and makes them 0 if they are negative using a similar concept as SIMD Within a Register (SWAR). After the activation layer, the Max Pooling layer reduces the feature dimensions, the number of parameters, and computations in the network using split x-y pooling, contributing to performance improvements while reducing the need for additional memory. The fully-

Table 6.2 Layer parameters for the CIFAR-10 CNN. Adapted from [6]

	Layer type	Filter shape	Output shape
1	Convolution	$5 \times 5 \times 3 \times 32$	$32 \times 32 \times 32$
2	Max Pooling	3×3	$16 \times 16 \times 32$
3	Convolution	$5 \times 5 \times 32 \times 32$	$16 \times 16 \times 32$
4	Max pooling	3×3	$8 \times 8 \times 32$
5	Convolution	$5 \times 5 \times 32 \times 64$	$8 \times 8 \times 64$
6	Max pooling	3×3	$4 \times 4 \times 64$
7	Fully-connected	$4 \times 4 \times 64 \times 10$	1×10

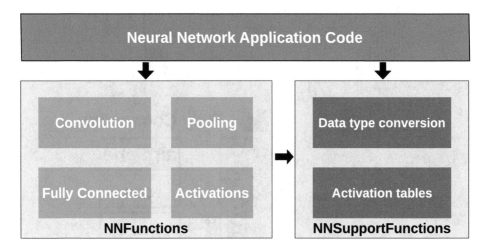

Fig. 6.2 Overview of the CMSIS-NN kernel structure. Adapted from [6]

connected layer has a matrix-vector multiplication that can also be implemented with a 1×2 kernel size to improve even more the execution time. This layer has full connections to all activations in the previous layer to define the output classification probabilities.

6.1.2 Soft Error Reliability Assessment of CIFAR-10 CNN Execution on Resource-Constrained IoT Devices

This Section aims to explore the erroneous behavior of the adopted CNN when exposing its layers to faults (i.e., flipped bits). Section 6.1.2.1 presents the early results when considering the execution of the CIFAR-10 CNN in two Arm Cortex-M processors: Cortex-M3 and Cortex-M4.

Table 6.3 Experimental setup

ML model	Cifar-10 CNN
API/Libraries	CMSIS-NN
Dataset	Cifar-10
Arm processors	Cortex-M3 (47 millions)
(executed instructions)	Cortex-M4 (15 millions)
Target FI	Register file, Function lifespan
CNN topology	3 Convolution (3 ReLU Activations), 3 Max pooling, 1 Fully-connected
Number of FI campaigns	22
Injections per campaign	17 k
Total fault injections	374 k

6.1.2.1 Experimental Setup

In order to estimate the percentage of errors from the CNN case study, the results are obtained by injecting bit-flips in the general-purpose registers (i.e., r0–r15) of both Cortex-M3 and Cortex-M4 Arm processors considering *random register file* and *function lifespan* FI techniques. Table 6.3 shows the experimental setup used to perform the proposed evaluation.

Each fault injection considers a single input image for one CNN execution, thus not considering fault propagation to the subsequent executions. One of the main concerns when assessing the soft error reliability of a system is to develop a precise, well-covered, and realistic approach. In this sense, this work sought to ensure that the number of fault injections has a statistical significance by applying the equations developed by [8]. These results comprise 17 k fault injections per campaign, thus generating a 1% error margin with a 99% confidence level. Although both processors share the same architecture (ARMv7-M) and instruction set (Thumb-2), the Cortex-M4 has an additional Digital Signal Processor (DSP) extension with a range of saturating and SIMD instructions. For that reason, Cortex-M3 requires $3\times$ as many instructions to execute the same CNN (Table 6.3). In addition, these results consider the same compilation environment: GCC 9.3 and optimization flag –O2.

6.1.2.2 CIFAR-10 CNN Execution Lifetime

Asserting the experimental setup accuracy is paramount to obtain meaningful and useful results. In this regard, the CIFAR-10 CNN was executed in the absence of faults over a RTL description of the Arm Cortex-M3, comparing gathered outputs with those obtained with the SOFIA execution. Results show that the CIFAR-10 CNN execution in SOFIA presents no difference in the output probabilities for the selected image w.r.t. the RTL execution (4.5 h), thus validating the utility of the adopted framework, which needs only 1 s to execute the

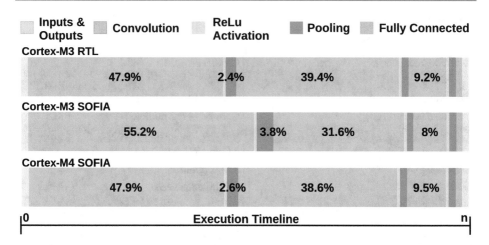

Fig. 6.3 CNN timeline with CMSIS-NN Kernel executing on Cortex-M3 and Cortex-M4 processors

same inference CNN model. Figure 6.3 shows the CNN execution detailing the lifespan of each layer, considering both the Arm Cortex-M3 and the Cortex-M4.

Observing the functions' lifespan in Fig. 6.3, it is notable that convolution layers execute at most of the time while the remaining layers execute in a small part of the CNN execution. In this sense, it is expected that faults injected randomly are more likely to reach the convolution layers, having a considerable impact on the CNN execution. The execution percentages for the RTL model are based on clock cycles, while in SOFIA, the percentages are based on the number of executed instructions. Such percents differ due to the number of required clock cycles to execute some instructions (e.g., division requires 2–12 cycles in Cortex-M3 RTL description). Note that the RTL description of the Cortex-M4 is not available at the Arm University program. These findings assist in the decision to properly evaluate the fault impact on the CNN layers considering both processor architectures.

6.1.2.3 CIFAR-10 CNN Soft Error Reliability Assessment

Figure 6.4 shows the results for the FI campaigns targeting all CNN layers. Such results are compared with those obtained from fault injection campaigns that target individual CNN layers and activation functions.

The CNN functions use several loops to process data, which requires many control instructions, consequently generating high UT and Hang occurrence (i.e., 22% on average) in both architectures. Even with a reduced number of crashes (UT and Hangs), ReLU and pooling layers show high soft error vulnerability in the Cortex-M3 when compared to Cortex-M4 (i.e., high occurrence of OMM and ONA). The results also show a higher masking effect on Cortex-M4 due to the CMSIS-NN optimizations targeting SIMD instructions (i.e., ~20% more vanished when comparing to Cortex-M3), reflecting such masking factor when

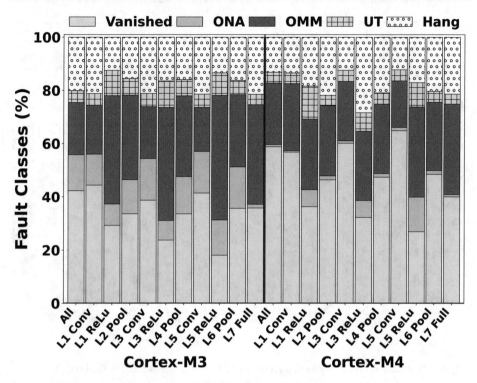

Fig. 6.4 Fault classification when injecting faults in the CNN layers

injecting faults on the CNN Convolution layers. In turn, the injections in the ReLU activation layers produced similar results in both processors. The ReLU layers perform a reduced data set (i.e., less execution time), thus reducing the masking rate even without SIMD optimizations.

The MWTF [9] metric is used to evaluate the CNN soft error reliability, since it shows the average amount of work an application can perform until reaching a failure (i.e., higher values are better). Figure 6.5 shows the MWTF of each layer normalized with the CNN MWTF for fault injections on the Cortex-M3. The results show a reasonable MWTF reduction regarding the occurrence of soft errors in the different layers. While soft errors from pooling layers reduce the MWTF in 35% on average, the faults from ReLU activation layers achieve an MWTF reduction of 62%. In this manner, the activation layers are critically affected in both architectures once such functions present high fault occurrence. Our findings denote that the CNN reliability is related to the layer types and its topology since the soft error occurrence increases from input to output layers.

Table 6.4 compares the output data from the fault injection campaigns along with the fault-free execution. The effective faults are very similar in both processors once the ONA fault types do not affect the output probabilities, thus not affecting the CNN accuracy. The

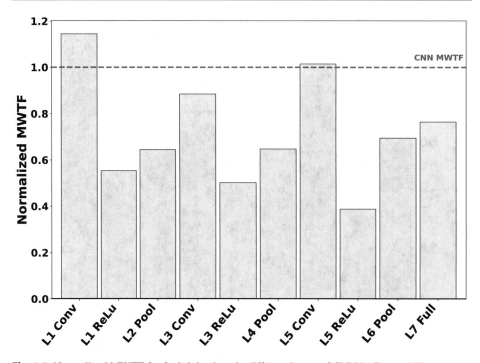

Fig. 6.5 Normalized MWTF for fault injections in different layers of CNN in Cortex-M3

Table 6.4 Percentages of tolerable and critical faults in the CNN

	Layers	Cortex-M3				Cortex-M4			
		Correct	Tolerable	Critical	Crash	Correct	Tolerable	Critical	Crash
1	Convlution	56.0	10.3	8.2	25.5	57.3	13.0	12.2	17.4
	ReLU	37.2	3.6	37.3	21.9	42.7	5.4	21.2	30.8
2	Max Pooling	46.5	21.7	10.1	21.8	47.9	17.8	8.6	25.7
3	Convlution	54.4	12.8	6.7	26.1	61.1	10.1	12.3	16.6
	ReLU	31.1	2.7	39.8	26.4	38.6	5.0	21.1	35.4
4	Max Pooling	47.6	21.0	9.4	22.0	48.8	17.6	8.5	25.1
5	Convlution	57.1	7.8	8.8	26.4	66.1	7.5	10.0	16.4
	ReLU	31.3	5.0	41.8	21.9	39.9	6.5	27.4	26.2
6	Max Pooling	51.2	18.8	8.4	21.6	49.9	18.7	7.0	24.4
7	Fully-Connected	37.2	15.4	22.1	25.4	40.9	11.4	22.6	25.1
All	All	55.8	11.8	7.8	24.5	59.6	11.1	12.0	17.2

effective and the critical faults achieve higher percentages while tolerable faults have lower ones. Note that highlighted results represent the worst-case scenarios. While pooling and convolution layers have high tolerable faults, the ReLU and the fully-connected functions are critically affected by the injected faults in both architectures. Even with tolerable faults,

Table 6.5 Experimental Setup

ML model	Cifar-10 CNN
API/Libraries	CMSIS-NN
Dataset	Cifar-10
Arm processor	Cortex-M3 (47 millions)
(executed instructions)	Cortex-M4 (15 millions)
Target FI	Register files, Flash (code and parameters), RAM
Number of FI campaigns	6
Injections per campaign	17 k
Total fault injections	102 k

critical faults reach from 13% up to 60, 85% of the output data, which reveals the vulnerability on the accuracy of the adopted CNN. Such results demonstrate that the occurrence of soft errors has a significant impact on the CNN execution and its accuracy when considering the individual analysis of distinct layers and processors.

6.1.3   The Impact of Soft Errors in Memory Units of Edge Devices Executing CIFAR-10 CNN

This Section explores the CIFAR-10 CNN's behaviour in the presence of soft errors by applying fault injections to the register file and isolated memory sections.

6.1.3.1 Experimental Setup

Table 6.5 shows the experimental setup, where 102k bit-flip are injected in the general-purpose register file (i.e., r0–r15) and Flash and RAM memory sections. This section applies the equations developed by Leveugle et al. [8] to ensure the statistical relevance of the fault injection results [10]. In this sense, each FI campaign includes 17 k experiments, providing results with an 1% error margin and 99% confidence level. Note that these experiments do not consider the propagation of faults to subsequent executions, as each experiment consists of one bit-flip injection and a single CNN execution with a single input image.

The FI campaigns are conducted considering both Arm Cortex-M3 and M4 processors. Although both Arm Cortex-M3 and M4 rely on the same instruction set architecture (ARMv7-M), results in Table 6.5 show that the Cortex-M3 requires 3× as many instructions w.r.t. to the M4. The reason for that is the DSP capabilities of Arm Cortex-M4, which includes saturating arithmetic and SIMD instructions. Table 6.6 shows that such ISA specific optimizations reflects on the memory occupation of the target evaluation boards.

Table 6.6 On-chip Memory Occupation (MO) for the inference CNN models, considering code, parameters and data

Arm	Reference chip	On-chip memory		CNN inference model			MO$_{(\%)}$
Processor		Flash$_{(kB)}$	SRAM$_{(kB)}$	Code$_{(kB)}$	Parameters$_{(kB)}$	Data$_{(kB)}$	
Cortrex-M3	STM32F103RCT6	256	48	4.3	36.5	43.2	27.63
Cortrex-M4	STM32F407ZGT6	1024	192	7.6	36.5	43.2	7.15

6.1.3.2 Soft Error Reliability Analysis

Figure 6.6 shows the CIFAR-10 CNN soft error results gathered from the injection of bit-flips in the register file, Flash and RAM memory sections considering the Arm Cortex-M3 and M4. The results show that the RAM section has no *crash* occurrence. This is because this type of fault does not affect the application's object code and NN parameters, but rather the CNN application data. In turn, the Flash section presents less *crash* occurrence than the register file. In regard to the memory sections, Flash and RAM show many *critical faults* because these faults are not overwritten at runtime (i.e., up to 2.3× more *critical faults* than the register file).

To understand the nature of the memory section faults, the CNN code, parameters and data are evaluated separately considering both Flash and RAM sections. Table 6.7 details the fault impact of the Flash code section for both Arm Cortex-M3 and M4 processors. Note that processor architecture influences the resulting CNN object code and therefore its vulnerability to bit-flips. The main difference relies on the use of SIMD instructions, i.e., the split of the convolution layers into convolution and matrix multiplication and the use of standard *memset* and *memmove* functions to fill out buffers in the Cortex-M4. These features directly impact on the occurrence of soft errors, since the variation in memory footprint and execution lifespan alter the vulnerability window (i.e., probability of a fault strike).

Table 6.7 also shows the fault percentages for each function object code section. Percentages are normalised according to the number of faults that affect each memory section, disregarding the faults that fall into portions of the code that do not influence the CNN execution (i.e., initialization and architecture-specific functions). These results show that the Arm Cortex-M4 is more reliable than Arm Cortex-M3. On the Arm Cortex-M3 side, activation and pooling are the most affected functions. In turn, pooling, fully-connected, and softmax present more soft errors when executed on the Arm Cortex-M4, because they have less SIMD instructions.

To understand the nature of the errors shown in Figs. 6.6, 6.7 presents the fault impact of the Flash memory addressing (i.e., CNN object code) for the two Arm processor architectures. The main difference between the object codes in these architectures is the use of SIMD instructions and the split of the convolution layers into convolution and matrix multiplication in Arm Cortex-M4. Considering these differences, the faults that occur in Arm Cortex-M3 have unaffected sections that comprise saturation instructions. In the Arm Cortex-M4, the non-affected sections also include these instructions along with code portions with SIMD

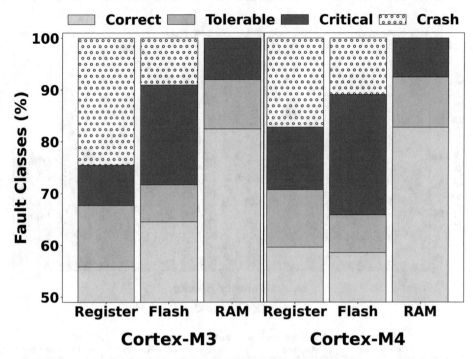

Fig. 6.6 Results showing fault classifications for Register File, Flash, and RAM memory sections for the CIFAR-10 CNN considering two Arm processors

Table 6.7 Percentages of memory occupation (MO), execution lifespan (ELS), and classification of faults occurring on CIFAR-10 CNN functions object code stored in Flash memory section

Sections	MO		ELS		Correct		Tolerable		Critical		Crash	
	M3	M4	M3	M4	M3	M4	M3	M4	M3	M4	M3	M4
Convolution	34.3	39.6	94.8	9.7	35.1	39.3	5.4	4.8	35.0	37.3	25.5	18.6
Matrix M.	–	21.2	–	85.6	–	52.0	–	5.4	–	25.1	–	17.5
ReLu	1.2	1.4	0.4	0.6	13.2	41.7	4.7	3.0	43.4	32.7	38.7	22.6
MaxPool	14.1	8.9	4.6	3.0	10.3	17.4	4.4	12.2	50.3	35.5	35.0	34.9
Fully C.	24.9	10.8	<0.1	0.1	24.1	28.2	14.7	9.3	33.6	37.2	27.5	25.3
Softmax	5.2	2.1	<0.1	<0.1	16.8	12.2	21.9	19.7	41.7	48.4	19.6	19.7
Main	20.3	9.7	0.2	0.2	51.5	52.0	4.8	4.3	29.1	28.0	14.6	15.7
MemFunc	–	6.3	–	0.8	–	58.3	–	1.9	–	15.6	–	24.2

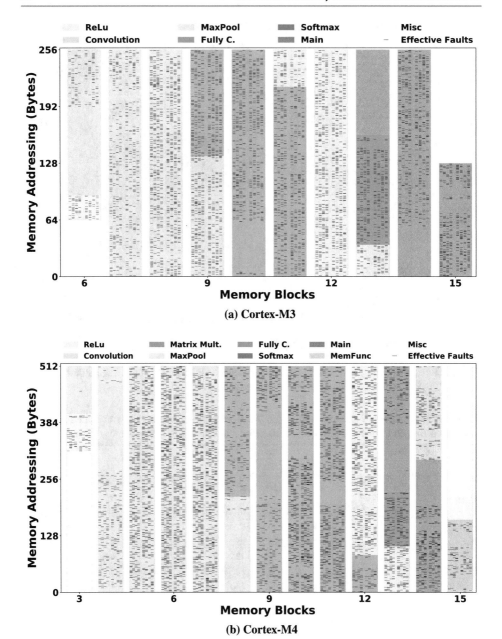

(a) Cortex-M3

(b) Cortex-M4

Fig. 6.7 Fault effects on Flash memory sections considering the CNN object code

instructions. In addition, the first portion of the convolution code object is flawless due to the understanding of the first layer and, consequently, it has a smaller execution window than the other CNN functions.

Aiming to analyse the soft error impact between Flash and RAM sections, Fig. 6.8 details the faults on the CNN parameters and data for the two Arm processor architectures. The most affected CNN parameters (i.e., Flash) are weights and bias, while the most affected data (i.e., RAM) is the scratch buffer. Note that the input figure (*InFig*) is not affected since the faults occur during the execution of the CNN. Image-to-column (*Im2Col*) is an auxiliary buffer used for matrix to column conversion, thus presenting either, high masking effect and low probability to propagate faults to the outputs. Weights and bias are key parameters used in CNN inferences, and therefore faults that occur in these areas tend to slightly impact on the output classification probabilities. In turn, the scratch buffer stores the inferences' results between layers, which present more faults. This is because the buffer is used as input and output of the convolution layers in a streaming fashion, consequently affecting the output probabilities.

In regard to the faults of the two processor architectures, the most significant difference was bias, which comprises essential data used in inferences. In this case, the number of instructions executed on Arm Cortex-M4 is $3\times$ less than on Arm Cortex-M3, opening a

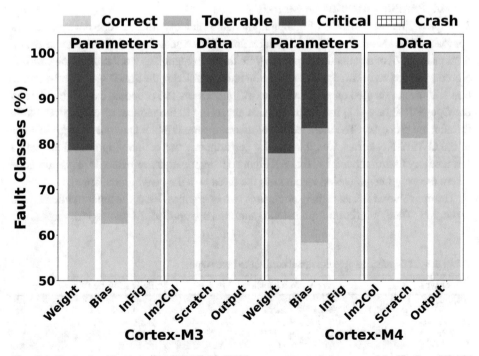

Fig. 6.8 Fault classifications for CIFAR-10 CNN parameters and data stored in Flash and RAM memory sections for Arm Cortex-M3 and M4 processors

vulnerability gap where the fault can occur before or during the inference, which might affect the output classification probabilities.

6.1.4   Impact of Thread Parallelism on the Soft Error Reliability of CIFAR-10 CNN

This Section proposes a multi-threaded version of the convolution neural network based on the CMSIS-NN kernel [6] applying to the Arm Cortex-A processor (i.e., ARMv7-A).

The first experiment analyses the CNN profile to understand the CNN application structure and which functions are most used. Table 6.8 shows the execution time percentages for a single thread running on the Arm Cortex-A7 processor according to each CNN layer kernel. These values demonstrate that the convolution layers execute almost all the time (i.e., 97%) in both the Thumb and SIMD instruction sets. Therefore, these layers are the best candidates to be parallelized to achieve such a performance improvement. In this regard, the following Section describes changes made to the CMSIS-NN kernel to allow parallel execution of convolution layers.

6.1.4.1 Parallel Convolution Kernels

To improve the performance of computational and memory limited edge devices, current approaches have been applying fixed-point arithmetic and quantization of weights and activations on 8-bit [6] or smaller data types [11]. On the other hand, this work extends the CMSIS-NN convolution kernel to support multiple threads considering the SIMD and Thumb instruction sets when executed on an Arm Cortex-A7 processor. The proposed extension relies on the OpenMP library [12] to support threads according to the number of cores available in the adopted processor. The main feature of this extension is that it does not change the goals of the CMSIS-NN kernel, i.e., it keeps the performance and memory optimizations with a low memory footprint overhead (<1%). Figure 6.9 demonstrates an example of pseudo-code where the proposed extension instantiates the threads in the convolution kernel.

The convolution kernel usually comprises two operations. First, the input reordering and expansion. Then, matrix multiplication operations are applied. In this regard, threads are

Table 6.8 CNN runtime percentages according to layer type

Layer type	Thumb runtime (%)	SIMD runtime (%)
Convolution	**97.15**	**97.28**
MaxPooling	2.59	2.22
ReLU	0.23	0.40
Fully-Connected	0.02	0.10

defined between the two operations not to change the CMSIS-NN kernel's characteristics. Furthermore, the *APP_THREADS* environment variable specifies the maximum number of threads to run in parallel, and the *omp_set_dynamic* method guarantees that the number of threads will be the same as defined by the user in the *num_threads* method, avoiding bottlenecks, as shown in Fig. 6.9. In addition, the *SIMD* and *ordered* directives indicate that the loop can be turned into a SIMD loop (i.e., multiple iterations of the loop can be executed concurrently using SIMD instructions), and those operations must be ordered to ensure they do not affect the inference precision.

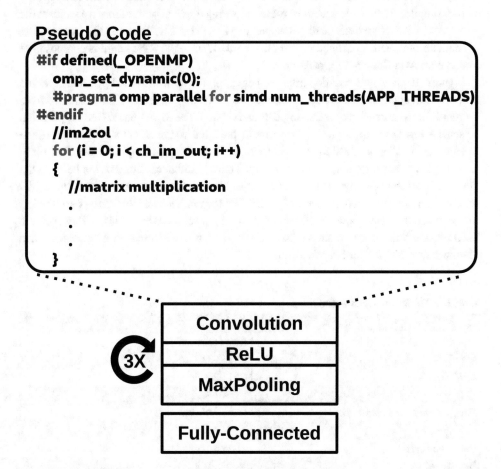

Fig. 6.9 CMSIS-NN extension to provide a parallel convolution kernel

6.1.4.2 Multi-thread Speedup Analysis Experimental Setup

The first experiment aims to reveal the speedup gains when applying thread parallelism to the CMSIS-NN convolution kernel. Table 6.9 shows the experimental setup used to evaluate the relative speedup considering the Thumb and SIMD instruction sets. To ensure results consistency, we consider the GCC compiler version 9.3 along with optimization flag $-O2$ and OpenMP version 4.5 for all experiments. The optimizations from the CMSIS-NN library are disable in Thumb versions by the architecture specific flag $-D\_ARM\_FEATURE\_DSP = 0$, ensuring that the object code is generated entirely by the compiler. To provide accurate results regarding application runtime, we validate the proposed CMSIS-NN extension on a Raspberry Pi 2 (RPi2) Model b board [13]. In this sense, the number of threads is defined ranging from 1 to 4, since the RPi2 board is composed of a quad-core Arm Cortex-A7 processor.

Figure 6.10 shows the speedup improvement applying multi-threaded parallelism in the adopted CNN application. Such results are calculated from the execution time obtained from the multi-thread execution over the single thread version. The SIMD-based CNN application presents a low speedup improvement in multi-threaded versions. On the other hand, the Thumb-based CNN application shows an almost linear speedup improvement (i.e., up to $3.5\times$ improvement w.r.t. single-core version). This performance difference is because the Thumb implementation (i.e., generic convolution) has no optimization, whereas the SIMD version promotes architecture-specific CMSIS-NN optimizations. In this sense, the compiler can optimize the object code of the Thumb instruction set more than SIMD. However, the SIMD versus Thumb curve shows that SIMD still has a performance $2.8\times$ higher than Thumb even with the multi-core speedup.

Table 6.9 Validation experimental setup

Evaluation board	Raspberry Pi 2 Model b
Processor	Arm cortex-A7
ML model	Cifar-10 CNN
API/Libraries	CMSIS-NN
Dataset	CIFAR-10
Compiler and optimization flag	GCC 9.3 with $-O2$
OpenMP version	4.5
Instruction sets	Thumb and SIMD
Parallel threads	1, 2, 3, and 4

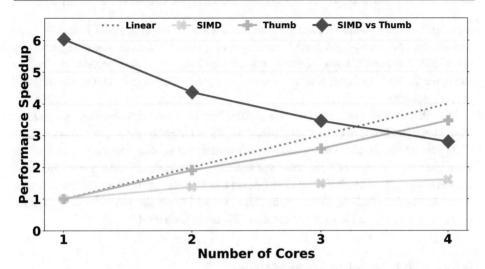

Fig. 6.10 Speedup results for 1, 2, 3 and 4 thread implementations for Thumb and SIMD instruction sets

6.1.4.3 Soft Error Reliability Analysis Experimental Setup

This Section explores the system soft error reliability by injecting faults (i.e., flipped bits) on processor registers during the execution of the CNN application with the proposed CMSIS-NN extension.

To assess the reliability of the proposed CMSIS-NN extension, we have used the SOFIA framework to inject faults in the general-purpose registers of the Arm Cortex-A7 processor (i.e., r0–r15). Note that all CMSIS-NN optimizations, including SIMD instructions and OpenMP extension, only require general-purpose registers. Table 6.10 shows the FI

Table 6.10 FI Experimental setup

Processor	Arm cortex-A7
ML model	Cifar-10 CNN
API/Libraries	CMSIS-NN
Dataset	CIFAR-10
Compiler and optimization flag	GCC 9.3 with $-O2$
OpenMP version	4.5
Instruction sets	Thumb and SIMD
Target FI	General-purpose registers
Number of FI campaigns	6
Injections per campaign	17 k
Total fault injections	102 k

experimental setup, which includes experiments with Thumb and SIMD instruction sets considering 1, 2, and 4 cores/threads running on the Arm Cortex-A7 model supported by the SOFIA framework. Such experimental setup considers the same version of compiler and optimization flags that was used for speedup analysis, keeping the consistency of the FI campaign results.

Note that each FI campaign contains N experiments, where each one contains a single input image and one CNN execution. Therefore, faults cannot be propagated to subsequent CNN executions. In addition, developing a realistic and precise approach is one of the main concerns for assessing the soft error reliability of a system. In this sense, this work applies the equations developed by Leveugle et al. [8]. These equations ensure the statistical significance of the fault injection results. Thus, each FI campaign has 17 k experiments to generate results with a 1% error margin and 99% confidence level.

6.1.4.4 Soft Error Reliability Assessment

Figure 6.11 presents the fault injection campaign results executing the adopted CNN application considering Thumb and SIMD instruction sets. The x-axis shows the Arm Cortex-A7 configurations for single, dual, and quad-core systems. The left y-axis shows the fault classes, while the red dots on the right y-axis show the MWTF gain compared to the single thread execution version.

The results show that soft errors can affect the Thumb version more than the optimised SIMD version in terms of crashes. These faults are mostly unexpected terminations (e.g., handling exceptions) and are faults detected by the architecture. Even when the number of threads is compared, the same behaviour is observed (i.e., similar percentages in single, dual, and quad-core). This is mainly due to a large number of control and memory instructions between the calculations. Furthermore, the CNN application runs on a Linux kernel; thus, the longer the execution time, the more kernel control instructions are executed.

On the other hand, the SIMD version exploits MAC instructions to provide optimizations in the inference phase, reducing the fault impact. In addition, CNN's performance improvement by increasing the number of threads also increases correct outputs and tolerable faults. However, the SIMD instruction set presented a slight increase in critical faults. This behaviour is caused by the more significant number of SIMD MAC instructions executed between load/store operations, which are more sensitive to soft errors affecting the registers that contain the inference data.

Finally, Fig. 6.11 also shows the impact of soft error reliability on parallel applications. Regarding the execution on dual and quad-core Arm Cortex-A7 processors, we can see that increasing the number of threads positively impacts the soft error reliability, reducing affected outputs (i.e., tolerable and critical faults) and increasing correct outputs for both SIMD and Thumb versions. Although the SIMD version shows a higher percentage of critical faults, it has the best trade-off between reliability and performance when comparing the normalised MWTF. This is due to the architecture-specific optimizations of CMSIS-NN

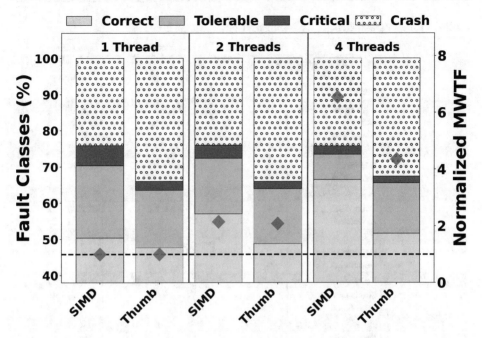

Fig. 6.11 Results from FI campaigns evaluating the thread parallelism with single, dual, and quad-core configurations

kernels aimed at SIMD instructions. According to the results shown in Fig. 6.10, even with an almost linear speedup in Thumb parallel execution, the SIMD parallel version still provides 3× more performance than the Thumb, consequently impacting the resulting MWTF.

6.2   Soft Error Reliability Assessment of the MobileNet CNN

This section presents the soft error reliability assessment results for the MobileNet CNN case study. First, Sect. 6.2.1 presents a detailed description of the MobileNet CNN developed with the CMix-NN library. Next, Sects. 6.2.2, 6.2.3, and 6.2.4 comprise the soft error reliability assessment considering the MobileNet CNN execution, topology, isolated layers, precision-bitwidth, memory (code, parameters and data), and soft error mitigation techniques.

6.2.1   MobileNet CNN Developed with CMix-NN

This Section presents the MobileNet case-study [14], which consists of a 29-layer CNN developed with the CMix-NN library [11] to run in SIMD featured Arm Cortex-M proces-

Table 6.11 Layer parameters for the Mobilenet CNN. Adapted from [14]

	Layer type	Stride	Filter shape	Input shape
1	Standard convolution	2	$3 \times 3 \times 3 \times 8$	$160 \times 160 \times 3$
2	Depthwise convolution	1	$3 \times 3 \times 8$	$80 \times 80 \times 8$
3	Poitwise convolution	1	$1 \times 1 \times 8 \times 16$	$80 \times 80 \times 8$
4	Depthwise convolution	2	$3 \times 3 \times 16$	$80 \times 80 \times 16$
5	Poitwise convolution	1	$1 \times 1 \times 16 \times 32$	$40 \times 40 \times 16$
6	Depthwise convolution	1	$3 \times 3 \times 32$	$40 \times 40 \times 32$
7	Poitwise convolution	1	$1 \times 1 \times 32 \times 32$	$40 \times 40 \times 32$
8	Depthwise convolution	2	$3 \times 3 \times 32$	$40 \times 40 \times 32$
9	Poitwise convolution	1	$1 \times 1 \times 32 \times 64$	$40 \times 40 \times 32$
10	Depthwise convolution	1	$3 \times 3 \times 64$	$20 \times 20 \times 64$
11	Poitwise convolution	1	$1 \times 1 \times 64 \times 64$	$20 \times 20 \times 64$
12	Depthwise convolution	2	$3 \times 3 \times 64$	$20 \times 20 \times 64$
13	Poitwise convolution	1	$1 \times 1 \times 64 \times 128$	$10 \times 10 \times 64$
14–23	Depthwise convolution	1	$3 \times 3 \times 128$	$10 \times 10 \times 128$
	Poitwise convolution	1	$1 \times 1 \times 128 \times 128$	$10 \times 10 \times 128$
24	Depthwise convolution	2	$3 \times 3 \times 128$	$10 \times 10 \times 128$
25	Poitwise convolution	1	$1 \times 1 \times 128 \times 256$	$5 \times 5 \times 128$
26	Depthwise convolution	2	$3 \times 3 \times 256$	$5 \times 5 \times 256$
27	Poitwise convolution	1	$1 \times 1 \times 256 \times 256$	$5 \times 5 \times 256$
28	Average pool	1	5×5	$5 \times 5 \times 256$
29	Fully-connected	1	256×1000	$1 \times 1 \times 256$
30	Softmax	1	Classifier	$1 \times 1 \times 1000$

sors. The CNN is trained with the ImageNet dataset [15], consisting of 10 million labeled images depicting 1000 object categories.

Table 6.11 shows the MobileNet layers that include multiple convolution layers interspersed by depthwise separable convolutions, with an average pooling and a fully-connected layer at the end of CNN. Moreover, Fig. 6.12 shows an overview of a typical CNN topology and the depthwise separable convolutions, which factorize a standard convolution into a depthwise convolution and a 1×1 pointwise convolution. In this factorization, depthwise separable convolution splits filtering and combining processes into two layers, drastically reducing computation and memory footprint.

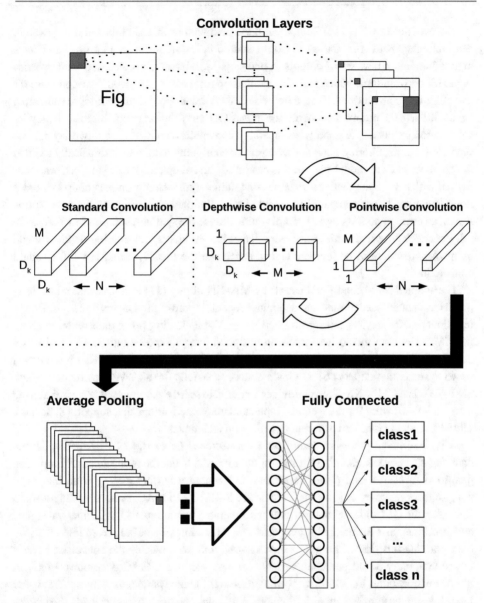

Fig. 6.12 Brief description of the MobileNet CNN topology. Adapted from [14]

The MobileNet CNN [14] is a streamlined architecture that aims to build lightweight deep inference networks. Different from standard convolution layers that filter and combine inputs into a new set of outputs in one step, MobileNet CNN is based on depthwise separable convolutions, which split this single step into two layers: depthwise convolutions and pointwise convolutions. First, depthwise convolutions apply a single filter to each input channel. Then, the pointwise convolution uses a 1×1 convolution to combine the depthwise convolution outputs. The depthwise separable convolution splits this into two layers, one separate layer for filtering and another layer for combining data, which drastically reduces the computation and model size. In this case study, the adopted MobileNet CNN was configured with a 3×3 depthwise separable convolution, representing savings of up to 9 times in computational cost compared to standard convolutions. In addition to the convolution layers, an average pooling layer is employed to reduce the spatial resolution to 1 before the fully connected layer. In this sense, MobileNet has 29 layers, considering depthwise and pointwise convolutions as separate layers, except for the first layer that is a full standard convolution.

The adopted MobileNet CNN uses the CMix-NN library [11] to implement mixed low-precision standard convolution and depthwise separable convolution layers functions. Unlike traditional CNN models that are trained using 32-bit floating-point data representation, the CMix-NN uses mixed low-precision unsigned integer representation. The CMix-NN kernels deploy optimizations focusing on enabling the execution of CNNs on Cortex-M based systems that support SIMD instructions, especially 16-bit *MAC* instructions (e.g., SMLAD). The CMix-NN library provides a complete set of convolutional kernels featuring a mixed low-bitwidth for the weights, input, and output activations that supports 8, 4, and 2 bitwidth combinations and different quantization techniques.

A typical mixed-precision *Quantized Convolutional Layer (QCL)* workload splits the convolution between quantized image-to-column and a matrix multiplication loop. The quantized functions load Q-bits input data in temporary buffers casting from the original Q-bits format to execute through vectorized SIMD 2×16 MAC instructions. Figure 6.13 illustrates the QCL internal components with memory requirements (e.g., inputs, weights, and structures) and the computational dataflow, which implement the mixed low-precision convolutional functions. The low-precision *MAC unit* accumulates the convolution result over a temporary 32-bit precision variable through vectorized MAC operations. In asymmetric quantizations, Z_w and Z_i apply the offset to the loaded parameter values to transpose them into the custom asymmetric domain. While the *Unpack* operation loads the convolution operands, the *Compressor* unit operates the final compression on the high-precision accumulation, considering a set of parameters T_A, which varies depending on the applied quantization technique.

In addition, CMix-NN library supports *Per-Layer (PL)* and *Per-Channel (PC)* compression techniques for any combination of bitwidth between input, output, and weights. While a PL quantization exploits a single *min/max* value for the entire layer, the PC computes a *min/max* value for any output channel. This latter approach is most beneficial when the

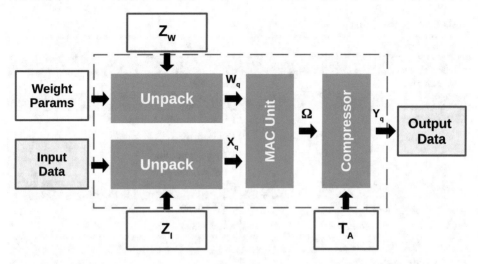

Fig. 6.13 CMix-NN quantized convolutional layer. Adapted from [11]

weight distribution varies widely between channels. Furthermore, the CMix-NN library also supports the *Inter-Channel Normalization (ICN)* activation [16]. This technique allows the introduction of lower bitwidth models with negligible inference loss, opening opportunities to exploit the soft error reliability of convolutional layers with precision bitwidth. The MobileNet mixed precision parameters have been acquired through the training scripts[1] provided by Rusci et al. [16].

6.2.2   The Impact of Precision Bitwidth on the Soft Error Reliability of the MobileNet CNN Execution on Resource-Constrained IoT Devices

This Section explores the soft error reliability of the MobileNet CNN considering different aspects: precision bitwidth and layer vulnerability analysis. In this sense, Sect. 6.2.2.1 details the experimental setup used to perform the fault injection campaigns. Section 6.2.2.2 rely on to analyze the execution lifetime of the adopted CNN. Furthermore, Sect. 6.2.2.3 exploits the MobileNet soft error reliability considering different precision bitwidth configurations.

6.2.2.1 Experimental Setup

To estimate the percentage of errors from the adopted case study, the two FI techniques mentioned in Sect. **??** are used to obtain the results, by injecting bit-flips in the general-purpose registers (i.e., r0–r15) of the Arm Cortex-M7 processor. Note that all optimizations

---

[1] https://github.com/marco-fariselli/training-mixed-precision-quantized-networks.

Table 6.12 Experimental setup

Arm processors	Cortex-M7
ML model	MobileNet CNN
API/Libraries	CMix-NN
Dataset	ImageNet
CNN topology	1 Standard convolution,
	13 Depthwise, 13 Pointwise,
	1 Average pooling,
	1 Fully-connected
Target FI	Register file, Function lifespan
Evaluated precisions	w4a4, w8a4,
(w: weights, a: activations)	w4a8, w8a8
Number of FI campaigns	40
Injections per campaign	17 k
Total fault injections	680 k

from CMix-NN, including SIMD instructions, requires only the general-purpose registers. Table 6.12 shows the experimental setup to perform the proposed evaluation. For this experimental setup we consider 8 and 4 bit precision quantizations in weights (i.e., w in Table 6.12) and input/output activations (i.e., a in Table 6.12).

In order to maintain the coherence, these results consider the same compilation environment: GCC 9.3 and optimization flag –O2. Each fault injection scenario considers a single input image for one CNN execution, thus not considering fault propagation to the subsequent executions. Developing a precise, well-covered, and realistic approach is one of the main concerns when assessing a system's soft error reliability. Aiming to ensure that the number of fault injections has a statistical significance, this work applies the equations developed by [8]. In this sense, the fault injection campaigns consider 17 k faults per scenario, thus generating a 1% error margin with a 99% confidence level.

6.2.2.2 MobileNet CNN Execution Lifetime

Asserting the experimental setup accuracy is paramount to obtain meaningful and useful results. To validate the adopted framework's utility, we compare the outputs reported in MobileNet[2] execution on the STM32H7 board with those gathered from the execution in the absence of faults over the SOFIA. The results from CNN execution in SOFIA presented no difference in the output probabilities for the selected image w.r.t. the on-chip execution. Figure 6.14 shows the MobileNet CNN execution detailing the lifespan of each layer, considering the execution on Arm Cortex-M7 through the STM32H7 board and SOFIA.

---

[2] https://github.com/EEESlab/mobilenet_v1_stm32_cmsis_nn.git.

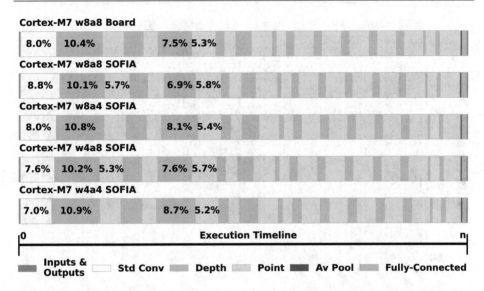

Fig. 6.14 CNN timeline with CMix-NN library executing on Arm Cortex-M7 processor

Observing the functions' lifespan in Fig. 6.14, it is notable that both standard convolution and depthwise separable convolution layers execute at most of the time while the remaining layers execute in a small part of the CNN execution. The variation of precision bitwidth also changes the function lifespans increasing the execution percent of depthwise separable convolutional layers. In this sense, it is expected that faults injected randomly are more likely to reach the convolution layers, having a considerable impact on the CNN reliability according to the precision bitwidth variations. Moreover, the execution percentages for the STM32H7 board are based on clock cycles, while in SOFIA, the percentages are based on the number of executed instructions. Such percents differ due to the number of required clock cycles to execute some instructions (e.g., a division might require from 2 to 12 cycles).

6.2.2.3 Soft Error Reliability Analysis

Aiming to evaluate the reliability considering the different aspects of the CNN execution, Fig. 6.15 shows the results for the FI campaigns targeting all CNN layers when considering different precision bitwidth configurations. Such results are compared with those obtained from fault injection campaigns that target distinct layers of the CNN topology considering different volumes of data processed. The CNN functions use several loops to process data, which requires many control instructions, consequently generating high UT and Hang occurrence (i.e., 15% on average) in all precision configurations. Fully-connected and pooling layers show high soft error vulnerability in all configurations since such layers do not vary in terms of quantized precision bitwidth. Observing the effect of quantizations on the CNN reliability, the quantization of inference weights affects reliability more than the quantization

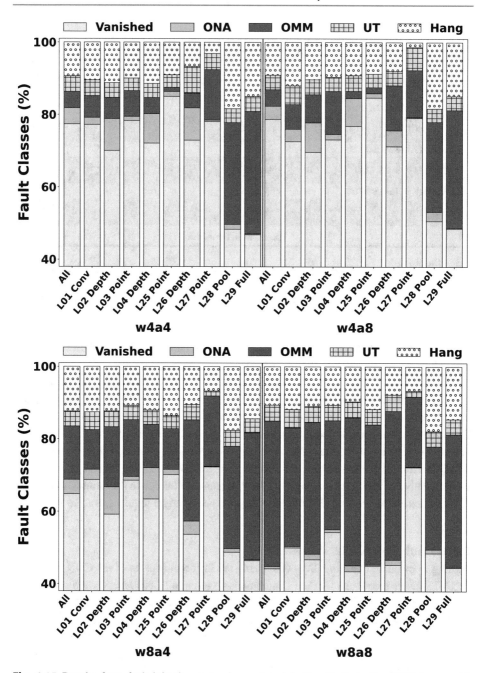

Fig. 6.15 Results from fault injection campaigns when injecting faults in the CNN layers with different precision bitwidth configurations

of activations since there is an increase in the masking rate when compared to *w8a8* (i.e., ~21% more vanished on *w4a4* and *w4a8* scenarios). According to the differences between the convolution layers, depthwise separable convolution layers are more susceptible to soft errors (i.e., high occurrence of ONA). Furthermore, the precision bitwidth can be linked to the increased vulnerability due to the greater probability of a failure spreading to the next layers (on average, OMM increases from 5% in *w4a4* to 35% in *w8a8* configurations).

Table 6.13 shows the results comparing the output data from the fault injection campaigns with the fault-free execution. These results allow us to evaluate the soft error impact on the CNN accuracy according to the classification presented in Sect. **??**. Considering the results of fault injections in all layers of the CNN, most of the effective faults tend to become critical, or system crashes in lower precision configurations (i.e., lower percents of tolerable faults in *w4a4*, *w4a8* and *w8a4*). Such an effect can be noted in the convolutional layers with

Table 6.13 Percentages of tolerable and critical faults in the CNN considering different precision bitwidth configurations

		w4a4				w4a8			
	Layers	Correct	Tolerable	Critical	Crash	Correct	Tolerable	Critical	Crash
1	Std. Convolution	79,16	0	5,99	14,84	76,03	0	6,76	17,21
2	Depthwise	78,87	0	5,79	15,34	77,79	0	7,74	14,47
3	Pointwise	79,46	0,03	7,03	13,48	74,58	0	11,86	13,55
4	Depthwise	80,21	0	4,44	15,35	84,51	0	1,97	13,52
25	Pointwise	86,32	0,83	0,32	12,53	85,85	1,10	0,55	12,49
26	Depthwise	81,85	3,83	0,19	14,13	75,68	11,48	0,91	11,93
27	Pointwise	78,48	10,36	3,56	7,59	79,34	11,26	1,70	7,70
28	Av. Pool	49,71	16,35	11,74	22,21	53,16	9,40	15,45	21,99
29	Fully-Connected	47,05	29,62	4,24	19,08	48,73	26,37	6,17	18,73
all	All	81,83	1,29	3,18	13,71	82,36	0,80	3,72	13,12
		w8a4				w8a8			
	Layers	Correct	Tolerable	Critical	Crash	Correct	Tolerable	Critical	Crash
1	Std. Convolution	71,59	0,04	10,85	17,52	50,38	24,50	8,19	16,94
2	Depthwise	66,68	0	16,61	16,71	48,21	31,28	5,09	15,41
3	Pointwise	69,64	0,04	15,56	14,77	55,01	24,13	5,93	14,93
4	Depthwise	72,06	0	11,90	16,04	45,05	35,57	5,26	14,12
25	Pointwise	71,61	10,64	0,55	17,21	45,26	32,93	5,78	16,04
26	Depthwise	57,29	26,36	1,58	14,76	46,64	32,44	8,66	12,26
27	Pointwise	72,42	14,90	4,56	8,11	72,42	15,32	3,98	8,28
28	Av. Pool	49,68	17,56	10,76	21,99	49,48	14,58	13,94	22,01
29	Fully-Connected	46,62	31,15	3,92	18,31	44,54	31,48	5,28	18,71
all	All	68,78	2,43	12,29	16,50	44,79	35,49	4,69	15,04

a high volume of data processing where the tolerable faults tend to 0% (i.e., layers 1–4). In turn, the number of effective faults increases when reducing the volume of data processed in the convolution layers (i.e., layers 25 and 26), where the number of effective faults increases when incrementing the precision bitwidth. Such results demonstrate that soft errors have a significant impact on the CNN execution and its accuracy, when considering the individual analysis of distinct layers and precision bitwidths.

6.2.3 The Impact of Soft Errors in Memory Units of Edge Devices Executing the MobileNet CNN

This Section explores the MobileNet CNN's behaviour in the presence of soft errors by applying fault injections to the register file and isolated memory sections.

6.2.3.1 Experimental Setup

Table 6.14 shows the experimental setup, where 459k bit-flip are injected in the general-purpose register file (i.e., r0–r15) and Flash and RAM memory sections. This work applies the equations developed by Leveugle et al. [8] to ensure the statistical relevance of the fault injection results [10]. In this sense, each FI campaign includes 17 k experiments, providing results with an 1% error margin and 99% confidence level. Note that these experiments do not consider the propagation of faults to subsequent executions, as each experiment consists of one bit-flip injection and a single CNN execution with a single input image.

The FI campaigns are conducted considering the Arm Cortex-M7. As mentioned before, the MobileNet was developed with the CMix-NN library, which only supports SIMD instruc-

Table 6.14 Experimental setup

Arm processor	Cortex-M7
(executed instructions)	(400 millions)
ML model	MobileNet CNN
API/Library	CMix-NN
Dataset	ImageNet
Precision bitwidths	8, 4, and 2
API/Libraries	CMix-NN
Target FI	Register files, Flash (code and parameters), RAM
Number of FI campaigns	27
Injections per campaign	17 k
Total fault injections	459 k

Table 6.15 Comparison of On-chip Memory Occupation (MO) for 8-bit precision inference CNN models, considering code, parameters and data

Arm	Reference chip	On-chip memory		CNN inference model			MO$_{(\%)}$
Processor		Flash$_{(kB)}$	SRAM$_{(kB)}$	Code$_{(kB)}$	Parameters$_{(kB)}$	Data$_{(kB)}$	
Cortrex-M3	STM32F103RCT6	256	48	4.3	36.5	43.2	27.63
Cortrex-M4	STM32F407ZGT6	1024	192	7.6	36.5	43.2	7.15
Cortrex-M7	STM32H743VIT6	2048	1024	20.2	1492.3	450.1	63.87

tions. However, due to the large volume of data inherent to weights and bias of MobileNet and the limited amount of memory available in STM32F407ZGT6 (Table 6.15), this model was deployed in the STM32H743VIT6 that includes a Cortex-M7.

6.2.3.2 Soft Error Reliability Analysis

This case study assesses the soft error reliability of the MobileNet CNN considering its precision bitwidth variation defined according to the CMix-NN library. Figure 6.16 shows soft error results obtained from the injection of bit-flips in the register file, Flash and RAM sections considering the Arm Cortex-M7. The Flash results present a large number of faults— most of them are tolerable. In addition, the percentage of critical faults remains similar for most precision bitwidths, the exception is the *w8a8* configuration. The results for the RAM section show that the higher precision (i.e., weights *w* and activations *a*) the more critical errors are expected.

Figure 6.17 presents the relative trade-off between the normalised Mean Work To Failure (MWTF) and memory-saving, considering different precision bitwidth configurations. The MWTF shows the average amount of work an application can perform until reaching a failure (i.e., higher values are better) [9]. Here, the memory-saving represents the amount of Flash/RAM not required by an implementation with regard to the highest and lowest requirements. This measure is based on the *.text* (Flash area for CNN code), *.data* (Flash area for CNN parameters) and *.bss* (RAM area for uninitialised variables) values reported by the compiler.

Figure 6.17a shows that lower MWTF values are obtained at reduced precision bitwidths for code (CM), parameters (PM) and data (DM). This occurs because the incidence of a bit-flip in a reduced precision configuration is more likely to modify the code or weights, which might ultimately impact on the CNN's functional behaviour or output classification. It is important to mention that the opposed happens in register files whereas faults are more often masked during the inference.

In regard to the memory-saving, results show that higher precision bitwidths (e.g., Fig. 6.17i) incur a larger amount of parameters and data, which lead to a larger vulnerability window. In this sense, the occurrence of a soft error might have a greater impact on the precision bitwidth variation of weights (from *w2* to *w8*), with an increase in the occur-

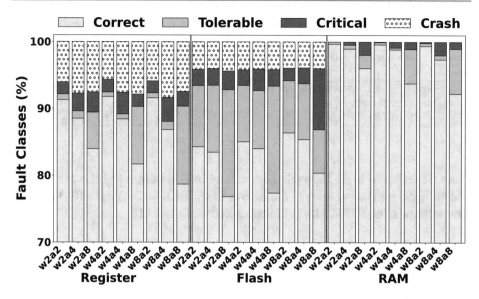

Fig. 6.16 Results showing fault classifications for Register File, Flash, and RAM memory sections for the MobileNet CNN according to the precision bitwidth

rence of critical failures when increasing the precision of activations (from *a2* to *a8*). This behavior is intensified in the *w8a8* configuration because, in this case, there is no precision offset between weights and activations and thus the fault tend to propagate through the CNN execution.

The results demonstrated that the soft error reliability of edge devices executing CNNs varies according to the affected memory section. If on the one hand the fault effects on register files are related to executed instructions and execution time. On the other hand, the fault effects on memory addressing are closely associated with the size of function object code, CNN parameters, and data. Furthermore, even in quantized solutions, a single bit-flip occurring on sensitive data stored in memory can compromise the entire CNN application.

6.2.4   Applying Lightweight Soft Error Mitigation Techniques to Embedded Mixed Precision Deep Neural Networks

This Section explores the soft error reliability of the MobileNet CNN considering different aspects: precision bitwidth, layer vulnerability, mitigation techniques, and relative trade-off analysis. In this sense, Sect. 6.2.4.1 details the experimental setup used to perform the fault injection campaigns. Section 6.2.4.2 exploits the MobileNet soft error reliability considering different precision bitwidth configurations. Then, Sect. 6.2.4.3 presents the reduction of soft errors when applying two system-level mitigation techniques. Finally, Sect. 6.2.4.4 presents

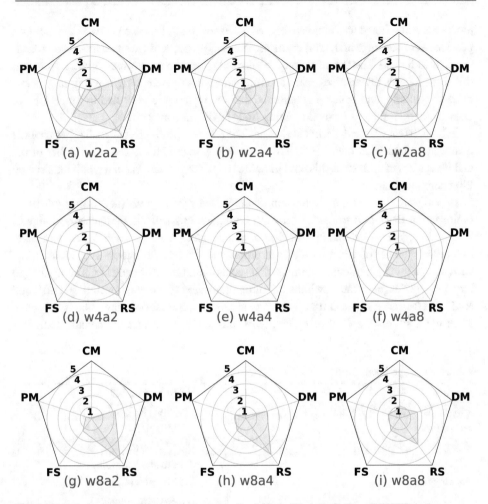

Fig. 6.17 Relative trade-off between the normalised *Mean Work To Failure* (MWTF) and memory-saving, considering different precision bitwidth configurations, comparing *Code MWTF* (CM), *Parameters MWTF* (PM), *Data MWTF* (DM), *Flash-Saving* (FS), and *RAM-Saving* (RS)

the relative performance and reliability trade-off for the adopted soft error mitigation techniques.

6.2.4.1 Experimental Setup

To provide trustworthy results, experiments consider more than 4.5 million fault injections to assess the soft error reliability of two mitigation techniques applied to the MobileNet on ImageNet. This work considers two FI techniques (i.e., *function lifespan* and *random register file*) to inject flipped bits in the general-purpose registers (i.e., r0–r15) of the Arm Cortex-M7 processor. This choice was motivated for two reasons. First, because the two FI techniques provide better coverage of the soft errors presented. Second, because all CMix-NN optimizations, including the SIMD instructions, require only the general-purpose registers. Note that this work focuses on the assessment and mitigation of soft errors originated from general-purpose registers, and hence it is assumed that the memory is protected by some type of error correction, such as Error Correcting Code (ECC) or parity bit.

Table 6.16 shows the experimental setup. The adopted MobileNet CNN has per-channel quantization with ICN layers (PC+ICN) [16], configured with the width multiplier of 0.5 and input sizes of 192, since this configuration has the minimum channel width required by 2-bit configurations.

In addition, this work uses the same compilation environment (i.e., *Clang 6.0.1* and optimization flag *–O2*) to set the bitwidth configurations and the mitigation techniques. Each fault injection campaign considers a single input image for each MobileNet CNN execution, thus not considering the fault propagation to the subsequent executions. Also, each campaign is a particular configuration scenario and the 270 campaigns comprise: 10 target layers * 9 precision bitwidths * 3 mitigation cases (Code unprotected, P-TMR, and RAT). Furthermore, conducting a precise, well-covered, and realistic approach is key when assessing a system's soft error reliability. In this sense, to ensure the results' statistical

Table 6.16 Experimental setup

Processor	Arm Cortex-M7
ML model	MobileNet CNN
API/Library	CMix-NN
Dataset	ImageNet
Topology	1 Standard convolution, 13 Depthwise, 13 Pointwise, 1 Average pooling, 1 Fully-connected
Target layers	All, L01, L02, L03, L04, L25, L26, L27, L28, L29
Evaluated precisions (w: weights, a: activations)	w2a2, w2a4, w2a8, w4a2, w4a4, w4a8, w8a2, w8a4, w8a8
Mitigation techniques	P-TMR and RAT
Number of FI campaigns	270
Injections per campaign	17 k
Total fault injections	4.59 millions

significance, this work injects 17 k faults per campaign, which according to [8], generates a margin of error of 1% with a 99% confidence level.

6.2.4.2 Soft Error Reliability Assessment of the MobileNet CNN Considering the Precision Bitwidth

Initially, we validated the SOFIA framework's faultless reference against the outputs reported by the MobileNet repository[3] and its on-chip execution. In this regard, we executed MobileNet CNN on an STM32H743[4] device, and then compared it with those collected from its execution on SOFIA, where no difference in the output probabilities was shown. This experiment is of paramount importance to guarantee the reproducibility and meaningfulness of the results.

After validating the reference flow, we generate fault injection campaigns considering the variation of precision bitwidth with weights raging from $w2$ to $w8$ and input/output activations from $a2$ to $a8$. Table 6.17 shows the MobileNet CNN soft error results detailing the fault classification for each bitwidth configuration.

In general, results show a similar soft error reliability behaviour between the different quantization configurations. For example, the results from fault injections in *All* layers in Table 6.17, crash occurrences handle 6.92% on average across all precision bitwidth configurations; this is because MobileNet CNN functions use multiple loops to process data that require many control instructions. Note that CNNs are known to have a lot of redundancy built-in, due to which they present a reasonable masking rate, which justifies the average of 86.7% of correct outputs. However, a single critical fault occurrence in safety-critical systems running the underlying trained models can lead to fatal consequences (i.e., life-threatening).

Table 6.17 also shows how the quantization of inference activations affects the MobileNet CNN soft error reliability by reducing the correct outputs in up to 10.12% when varying from $a2$ to $a8$. Although it affects less, increasing the weight bitwidth also reduces the soft error reliability by up to 2.21% on the correct outputs when varying the configurations from $w2$ to $w8$. This occurs because a SIMD MAC instruction splits a 32-bits register into different segments, which are set according to the bitwidth configuration. This leads to a higher probability to overwrite the faulty bit, thus reducing the probability to propagate the fault to the inference phases. Results show that both higher precision bitwidths and the unpack/compress process can led to an increase number of faults. The unpack/compress process is related to load and store instructions that are executed before and after SIMD MAC operations, which increases the probability of a fault impacts on the output probabilities. Note that the increase in the precision bitwitdh can also reduce the fault criticality since a soft error in a less significant bit is less likely to generate a critical fault. Such behaviour

---

[3] https://github.com/EEESlab/mobilenet_v1_stm32_cmsis_nn.git.

[4] https://www.st.com/en/microcontrollers-microprocessors/stm32h743vi.html.

Table 6.17 Percentages of Correct (CO), Tolerable (TO), Critical (CT) and Crash (CS) faults on MobileNet CNN considering different precision bitwidth configurations

		w2a2				w2a4				w2a8			
	Layers	CO	TO	CT	CS	CO	TO	CT	CS	CO	TO	CT	CS
1	Std Conv	91.65	0.01	2.03	6.31	89.98	0.01	1.42	8.58	89.81	0.41	0.93	8.85
2	Depthwise	92.66	0.08	0.15	7.11	92.77	0.01	0.24	6.98	85.62	5.02	2.95	6.41
3	Pointwise	90.58	0.07	3.76	5.58	88.32	0.12	3.44	8.12	83.31	2.06	6.23	8.40
4	Depthwise	91.61	0.06	1.55	6.77	88.55	0.00	4.37	7.08	79.93	6.59	7.11	6.36
25	Pointwise	92.84	1.31	0.47	5.39	89.75	1.85	0.58	7.82	87.53	3.66	0.78	8.03
26	Depthwise	84.49	7.10	1.29	7.12	84.44	7.13	1.26	7.17	82.22	11.27	0.55	5.96
27	Pointwise	80.04	9.84	2.13	8.00	79.94	11.15	1.21	7.70	79.28	12.19	0.71	7.81
28	Avg Pool	65.28	20.11	3.86	10.74	65.19	20.02	4.04	10.75	65.55	20.19	3.47	10.78
29	Fully-Con	75.08	17.57	1.89	5.46	74.30	18.15	1.93	5.62	74.19	18.17	1.97	5.67
All	All	91.32	0.84	1.80	6.04	88.51	1.11	2.66	7.72	83.96	5.45	3.02	7.56

		w4a2				w4a4				w4a8			
	Layers	CO	TO	CT	CS	CO	TO	CT	CS	CO	TO	CT	CS
1	Std Conv	91.81	0.00	1.75	6.45	90.18	0.01	1.08	8.74	89.60	0.60	0.87	8.93
2	Depthwise	92.84	0.00	0.15	7.01	92.79	0.05	0.18	6.98	84.63	8.72	0.65	6.00
3	Pointwise	90.64	0.03	3.78	5.55	88.51	0.01	3.47	8.01	81.47	6.35	4.31	7.86
4	Depthwise	91.60	0.00	1.54	6.86	88.04	0.04	5.03	6.89	78.94	13.78	1.19	6.08
25	Pointwise	92.11	2.04	0.25	5.60	89.95	1.92	0.42	7.71	85.76	5.78	0.54	7.93
26	Depthwise	85.49	6.77	1.21	6.53	84.36	7.09	1.50	7.05	80.51	13.05	0.46	5.98
27	Pointwise	81.51	8.35	2.32	7.83	82.10	8.89	1.43	7.58	80.22	10.81	0.94	8.04
28	Avg Pool	65.66	19.55	3.85	10.94	65.41	20.21	3.79	10.59	66.04	19.54	3.49	10.94
29	Fully-Con	75.19	17.18	1.95	5.67	74.21	18.16	1.99	5.64	74.42	17.91	2.16	5.51
All	All	91.79	0.66	1.88	5.67	88.43	0.72	3.24	7.61	81.68	8.61	1.86	7.85

		w8a2				w8a4				w8a8			
	Layers	CO	TO	CT	CS	CO	TO	CT	CS	CO	TO	CT	CS
1	Std Conv	92.88	0.01	0.70	6.41	90.19	0.01	0.87	8.93	89.78	0.64	0.66	8.92
2	Depthwise	92.85	0.06	0.19	6.89	92.06	0.03	0.32	7.58	83.90	8.90	0.54	6.66
3	Pointwise	90.56	0.04	3.58	5.82	87.73	0.02	3.87	8.38	78.65	11.58	1.49	8.28
4	Depthwise	91.60	0.06	1.39	6.95	86.15	0.02	6.35	7.48	75.77	17.18	0.68	6.36
25	Pointwise	89.99	4.14	0.20	5.68	88.72	3.23	0.28	7.77	79.27	9.22	3.54	7.97
26	Depthwise	84.19	7.03	1.62	7.16	82.93	7.94	1.92	7.21	79.98	8.45	4.94	6.64
27	Pointwise	81.24	8.51	2.49	7.76	81.39	8.86	1.64	8.11	84.01	6.58	1.64	7.78
28	Avg Pool	65.80	19.73	3.55	10.92	65.65	19.56	3.69	11.10	66.73	6.89	15.44	10.94
29	Fully-Con	74.52	17.67	2.13	5.68	73.56	18.39	2.11	5.94	75.79	16.56	2.23	5.42
All	All	91.60	0.74	1.78	5.88	86.84	1.22	3.61	8.34	78.71	11.64	2.21	7.45

can be seen in Table 6.17 *All* when comparing *w8a4* and *w8a8* configurations, where the number of critical faults decreases and tolerable faults increases significantly.

To understand the layers' vulnerability to the soft errors, Table 6.17 shows the results obtained from FI campaigns that target distinct layers of the MobileNet CNN topology considering different data volumes. Results show that the most effective faults tend to become either critical or system crashes in low-precision activation configurations (i.e., *a2* and *a4*). Such an effect can be seen in convolutional layers with a high volume of input data processing, where tolerable faults tend to 0% (i.e., layers 1 to 4). In turn, the number of tolerable faults increases as the input data volume reduces in the convolution layers (i.e., layers 25 and 27). This is because the dimensions of input activations are reduced during MobileNet CNN execution, while the channel width increases, thus making the data volume larger in weights w.r.t. input activations. Consequently, faults occurring in these layers are more likely to propagate to the output, i.e., the appearance of a high number of tolerable and critical faults in layers 26 and 27. Note that the occurrence of faults (i.e., Tolerable + Critical + Crash) also increases alongside the precision bitwidth.

6.2.4.3 Applying Mitigation Techniques to MobileNet CNN

Aiming to reduce the MobileNet CNN susceptibility to soft errors, this Section considers the use of two software-based mitigation techniques: P-TMR and RAT. Both techniques are applied to the matrix multiplication function, which is considered here the most critical one due to its higher active period within the MobileNet execution time.

Figure 6.18 shows the reliability improvement of MobileNet CNN by applying the two mitigation techniques. The x-axis has three bars for each adopted precision bitwidth configuration. The first bar represents MobileNet CNN with unprotected code (λ), and the other two the mitigation techniques P-TMR (γ) and RAT (β). The left-hand y-axis shows the soft error percentage obtained from the fault injection campaigns, and the right-hand y-axis

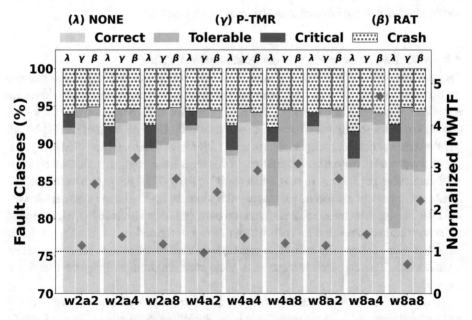

Fig. 6.18 Results showing fault classifications comparing MobileNet CNN without protection, with P-TMR, and RAT mitigation techniques. The red dots indicate the normalized MWTF (right y-axis)

shows the MWTF normalized by the unprotected version. While the left-hand metric shows an overview of the generated faults, the one at the right-hand side relies on a well-accepted reliability metric to compare the two mitigation techniques.

As expected, Fig. 6.18 shows that both mitigation techniques significantly increase the number of correct outputs while reducing the number of critical faults. In general, a lower precision bitwidth configuration (i.e., 2 and 4-bits to w and a) lead to a reduction in fault occurrences. On the other hand, 8-bit configurations present a higher occurrence of tolerable faults. This is due to larger bitwidth operations that reduce the fault masking rate. Even under these conditions, both mitigation techniques protect the code and turn critical faults into correct or tolerable ones. Figure 6.18 shows that P-TMR has a significant AVF improvement in all scenarios, but the performance penalty does not compensate for w4a2 and w8a8 configurations (i.e., normalized $MWTF < 1$). In turn, the RAT improved the MWTF in all configurations, raising up to $4.7\times$ in the *w8a4* precision bitwidth.

Figure 6.19 shows the reliability improvement per layer for the most affected precision bitwidth configurations (i.e., *w8a4* and *w8a8*). Compared to the unprotected version, both mitigation techniques show soft error reliability improvements. On the one hand, Fig. 6.19a shows a reduction of up to 8% in critical faults and system crashes in the 4-bit precision activation. On the other hand, Fig. 6.19b illustrates that some tolerable, critical, and crash faults become correct outputs for 8-bit precision activations. In the P-TMR perspective, this effect occurs mainly due to faults striking registers used by redundant instructions. Unlike, RAT reduces the number of vulnerable registers during the critical function's execution, taking advantage of the inherent high resilience of MobileNet.

6.2.4.4 Trade-Off Between Performance and Reliability

This Section details the drawbacks introduced by the two mitigation techniques and discusses the trade-off between increased protection and performance penalty. Figure 6.20 shows the performance overhead of the P-TMR and RAT compared to no protection execution. The execution times were extracted by running the MobileNet CNN on an STM32H743 board. Results show a performance degradation of up to $1.2\times$ for RAT and $3.8\times$ for P-TMR, depending on the precision bitwidth configuration. In this regard, the original MobileNet achieves ~ 1.9 inferences per second. In turn, when applying the P-TMR mitigation technique, the number of inferences per second reduces to ~ 0.5 while the RAT ~ 1.5 in worst-case scenarios. Note that the most remarkable performance overhead occurs because P-TMR is applied to the application's intermediate code without further optimization, i.e., the application is compiled with *−O2* and the mitigation technique is applied without architecture-specific optimizations. This approach is required to avoid code removals made by the compiler's backend.

Figure 6.21 compares the relative trade-off between reliability, accuracy, performance and memory footprint overhead for two precision bitwidth configurations (*w8a4* and *w8a8*). Table 6.18 shows the footprint overhead, which is calculated based on the additional hardened

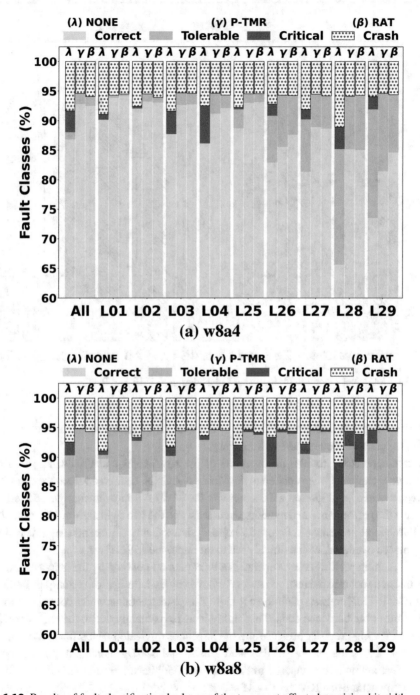

Fig. 6.19 Results of fault classification by layer of the two most affected precision bitwidth configurations

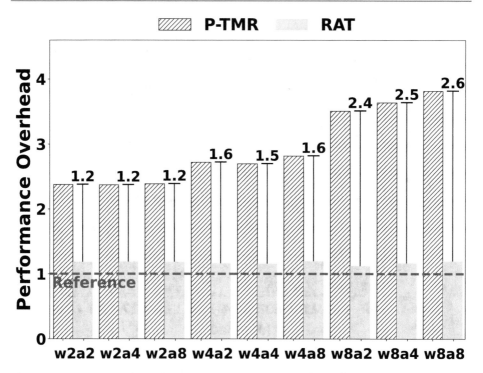

Fig. 6.20 MobileNet execution time overhead considering P-TMR and RAT

application code size resulting from both P-TMR and RAT mitigation techniques w.r.t. the original application code (i.e., Flash memory).

This comparison provides an overview of the advantages and disadvantages of both mitigation techniques when applied to the MobileNet CNN. Gathered values are normalized between 1 and 5, and the top axis represents the MWTF improvement, the two left axes represent the performance and memory overheads, and the two right axes show the precision and the tolerable percentages. Figure 6.21 clearly shows that RAT presents a significant lower performance overhead, which directly led to an improved MWTF w.r.t. the P-TMR.

The resulting performance overhead can be explained not only by the increased number of instructions but also the instruction set employed by each mitigation technique. Table 6.19 shows that P-TMR consists of almost 5× more *Thumb* instructions, which correspond to near 90% of the entire hardened code. This highly increase is mainly due to the register spilling forced by the high register pressure and the impact of replicated instructions. In turn, RAT slightly increases the number of executed *Thumb* and *SIMD*, maintaining the percentage of both instruction sets compared to the reference MobileNet execution. Aforementioned results demonstrate that RAT provides the best relative performance, reliability and memory footprint utilisation trade-off.

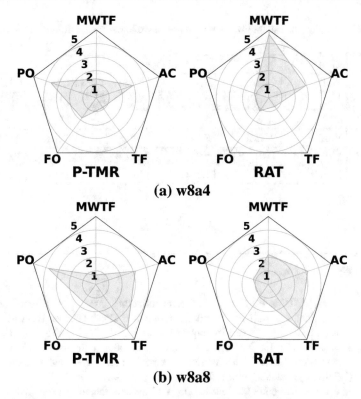

Fig. 6.21 Relative trade-off between P-TMR and RAT mitigation techniques considering *w8a4* and *w8a8* precision bitwidth configurations, comparing *Mean Work To Failure* (MWTF), *Performance Overhead* (PO), *Footprint Overhead* (FO), *Accuracy* (AC), and *Tolerable Faults* (TF)

Table 6.18 Normalized MobileNet footprint overhead when applying soft error mitigation techniques

Mitigation	w2a2	w2a4	w2a8	w4a2	w4a4	w4a8	w8a2	w8a4	w8a8
P-TMR	1.78	1.80	1.59	1.80	1.82	1.60	1.81	1.82	1.59
RAT	1.16	1.16	1.13	1.16	1.17	1.13	1.16	1.17	1.13

These results are of paramount importance for safety-critical applications because, in addition to reliability, these applications have real-time requirements. For example, in self-driving cars, a late reaction can lead to a fatal accident [17]. In this context, traditional soft error mitigation solutions involving time redundancy, such as Triple Modular Redundancy (TMR), may not be suitable for such kind of applications. Even when partially applied, this technique might inflict a significant response time penalty that is not tolerable in real-time applications, thus justifying the need for lightweight techniques, such as RAT, especially for resource constraint systems.

Table 6.19 Percentage of use and relative increase in executed instructions considering different instruction sets

	w8a4			w8a8		
Type	None	P-TMR	RAT	None	P-TMR	RAT
SIMD (%)	34.3	9.8	32.2	36.7	10.1	30.5
Thumb (%)	65.7	90.2	67.8	63.3	89.9	69.5
Type	None	P-TMR	RAT	None	P-TMR	RAT
SIMD	1×	1.1×	1.05×	1×	1.1×	1.06×
Thumb	1×	4.8×	1.15×	1×	4.9×	1.23×

References

1. Abich, G., Gava, J., Reis, R., Ost, L.: Soft error reliability assessment of neural networks on resource-constrained IoT devices. In: IEEE International Conference on Electronics, Circuits and Systems (ICECS), pp. 1–4 (2020). https://doi.org/10.1109/ICECS49266.2020.9294951
2. Abich, G., Reis, R., Ost, L.: The impact of precision bitwidth on the soft error reliability of the mobilenet network. In: IEEE Latin American Symposium on Circuits and Systems (LASCAS), pp. 1–4 (2020). https://doi.org/10.1109/LASCAS51355.2021.9667153
3. Abich, G., Gava, J., Garibotti, R., Reis, R., Ost, L.: Impact of thread parallelism on the soft error reliability of convolution neural networks. In: IEEE Latin American Symposium on Circuits and Systems (LASCAS), pp. 1–4 (2022). https://doi.org/10.1109/LASCAS53948.2022.9789088
4. Abich, G., Gava, J., Garibotti, R., Reis, R., Ost, L.: Applying lightweight soft error mitigation techniques to embedded mixed precision deep neural networks. IEEE Trans. Circuits Syst. I: Regul. Pap. **68**(11), 4772–4782 (2021). https://doi.org/10.1109/TCSI.2021.3097981
5. Abich, G., Garibotti, R., Reis, R., Ost, L.: The impact of soft errors in memory units of edge devices executing convolutional neural networks. IEEE Trans. Circuits Syst. II: Express Briefs **69**(3), 679–683 (2022). https://doi.org/10.1109/TCSII.2022.3141243
6. Lai, L., Suda, N., Chandra, V.: CMSIS-NN: efficient neural network kernels for arm cortex-M CPUS (2018). arXiv:1801.06601. https://doi.org/10.48550/arXiv.1801.06601
7. Krizhevsky, A., Hinton, G., et al.: CIFAR-10/100—learning multiple layers of features from tiny images. Technical report, University of Toronto (2009). http://citeseerx.ist.psu.edu/viewdoc/download?doi=10.1.1.222.9220&rep=rep1&type=pdf
8. Leveugle, R., Calvez, A., Maistri, P., Vanhauwaert, P.: Statistical fault injection: quantified error and confidence. In: Design, Automation and Test in Europe Conference (DATE), pp. 502–506 (2009). https://doi.org/10.1109/DATE.2009.5090716
9. Reis, G.A., Chang, J., et al.: Software-controlled fault tolerance. ACM Trans. Arch. Code Optim. (TACO) (2005). https://doi.org/10.1145/1113841.1113843
10. Krzywinski, M., Altman, N.: Points of significance: significance, p values and t-tests. Nat. Methods **10**(11), 1041–1042 (2013). https://doi.org/10.1038/nmeth.2698
11. Capotondi, A., Rusci, M., Fariselli, M., Benini, L.: CMIX-NN: mixed low-precision CNN library for memory-constrained edge devices. IEEE Trans. Circuits Syst. II: Express Briefs **67**(5), 871–875 (2020). https://doi.org/10.1109/TCSII.2020.2983648

12. Dagum, L., Menon, R.: OpenMP: an industry standard API for shared-memory programming. IEEE Comput. Sci. Eng. **5**(1), 46–55 (1998). https://doi.org/10.1109/99.660313
13. Pi, R.: Raspberry Pi 2 Model b (2021). https://www.raspberrypi.org
14. Howard, A.G., Zhu, M., Chen, B., Kalenichenko, D., Wang, W., Weyand, T., Andreetto, M., Adam, H.: MobileNets: efficient convolutional neural networks for mobile vision applications (2017). arXiv:1704.04861, https://doi.org/10.48550/arXiv.1704.04861
15. Deng, J., Dong, W., Socher, R., Li, L., Kai Li, Li Fei-Fei: ImageNet: a large-scale hierarchical image database. In: 2009 IEEE Conference on Computer Vision and Pattern Recognition, pp. 248–255. IEEE (2009). https://doi.org/10.1109/CVPR.2009.5206848
16. Rusci, M., Capotondi, A., Benini, L.: Memory-driven mixed low precision quantization for enabling deep network inference on microcontrollers (2019). arXiv:1905.13082. https://doi.org/10.48550/arXiv.1905.13082
17. Banerjee, S.S., Jha, S., Cyriac, J., Kalbarczyk, Z.T., Iyer, R.K.: Hands off the wheel in autonomous vehicles?: a systems perspective on over a million miles of field data. In: IEEE/IFIP International Conference on Dependable Systems and Networks (DSN), pp. 586–597 (2018). https://doi.org/10.1109/DSN.2018.00066

Conclusions and Future Work

7

The first goal of this Book comprises to make an extensive and statistical significance soft error consistency assessment of a JIT-based fault injection framework. This work has investigated the soft error assessment consistency of a JIT virtual platform simulator (SOFIA) with more than 12 million fault injections considering single and multi-core Arm processor architectures. The fault injection campaigns considered different cross-compilers, software stacks, programming models, and 52 applications. Results demonstrated that the architectural difference, such as the ISA, between the two Arm processors, affects the reliability of single-core systems. However, the addition of tiny operating systems (e.g., FreeRTOS) in the software stack did not affect the consistency of soft error assessment. Regarding cross-compilers, those based on LLVM appeared to be more reliable ones, with the best compiler set being the *Clang 6.0.1* using the *02* optimization flag.

Furthermore, we showed a worsened of the mismatch while increasing the number of cores that can be mitigated a little by improving the workload balance and context switching in parallel applications, such as in the MPI programming model. Finally, we demonstrated that by tuning the SOFIA for a more detailed simulation by the quantum size parameter (i.e., 44-instruction block), we obtained the best cost-benefit in terms of soft error assessment accuracy, with a worst-case mismatch of 8.76% and high simulation performance, reaching up to 345 MIPS. We conclude that achieved mismatches are acceptable and are not a hindrance to evaluate soft errors at early design phases. Further, given the remarkably achieved speedup, SOFIA's utilization appears promising since it can also be used to compare different processor models, ISAs, kernel, and complex benchmarks with billion of instructions. Finally, authors also believe that the high statistical significance presented gives to this work the potential to be a reference for other studies with concerns about soft error resilience of Arm processors. In this sense, the obtained results validates *the first hypothesis of this Book*. Considering the consistency of these results, it is possible to state that SOFIA provides a consistent soft error reliability assessment while achieving some performance

G. Abich et al., *Early Soft Error Reliability Assessment of Convolutional Neural Networks Executing on Resource-Constrained IoT Edge Devices*, Synthesis Lectures on Engineering, Science, and Technology, https://doi.org/10.1007/978-3-031-18599-1_7

w.r.t. RTL and gem5 simulators. Also, these findings allow us to proceed to assess the reliability of more complex benchmarks such as ML applications targeting resource-constrained IoT devices.

In the second goal of this Book, we use the SOFIA framework to assess the soft error reliability of CNNs developed based on CMSIS-NN and CMix-NN specialized libraries, which enable the execution of complex CNNs in Arm Cortex-M processor architectures featuring SIMD instructions. SOFIA's flexibility allows us to cover different aspects of a given software stack while adopting different fault injection techniques. This work uses two fault injection techniques (*Random register file* and *Function Lifespan*) to isolate specific moments of CNN's execution, making it possible to perform fault injections on each CNN layer. The first case study comprises the soft error reliability evaluation of the quantized CIFAR-10 CNN (i.e., 8-bit precision) developed with CMSIS-NN library considering two Arm Cortex-M processor architectures and trained with the CIFAR-10. The evaluated results demonstrate that the adopted CNN has a high susceptibility to soft errors since, in most cases, the effective faults exceed 50%. The activation layers are more susceptible to soft errors since critical failures affect both the MWTF and the accuracy of CNN. Furthermore, the results from Cortex-M4 show that SIMD based optimizations reduce the CNN susceptibility to soft errors when comparing to Cortex-M3. The second case study comprise the MobileNet CNN, which is developed based on the mixed low-precision CMix-NN library considering the Arm Cortex-M7 processor architecture and trained on the ImageNet dataset. The evaluated results demonstrate that the adopted CNN has a high susceptibility to soft errors in higher precision bitwidth configurations since, in most cases, the effective faults exceed 50%. Also, the precision bitwidth of the weights is more susceptible to soft errors than the activations, since critical failures affect both the reliability and the accuracy of CNN. Moreover, reducing the precision bitwidth of weights and activations further affects CNN's soft error reliability by increasing the masking rate by 21%. Considering such early evaluations, our third case study relies on apply lightweight soft error mitigation techniques to mixed precision MobileNet CNN. The evaluated results demonstrate that the adopted CNN has a high susceptibility to soft errors in higher precision bitwidth configurations, since the fault occurrence achieves up to 20%. Results also demonstrate that the variation of precision bitwidth of the activations is more susceptible to soft errors than the weights, since critical failures affect both the reliability and the accuracy of CNN. Moreover, the reduction of weights and activations precision bitwidth increases the fault-masking capability of up to 10%, thus reducing the MobileNet CNN susceptibility to the occurrence of soft errors. However, such precision bitwidth reduction does not eliminate the occurrence of critical failures, requiring the use of fault mitigation techniques. Finally, gathered results show that RAT provides significant soft error reliability improvement at a lower performance penalty (i.e., $\sim 1.2\times$ on average) when compared to the P-TMR. Such conducted studies validates *the second hypothesis of this Book*, which early investigates and identifies the correlation between FI results, NN optimized kernels, and reduced precision parameters of CNNs executing on resource-constrained IoT edge devices.

The third goal of this Book relies on investigate the soft error reliability of two CNN inference models deployed in three low-power Arm microprocessors. The results obtained from more than 500k fault injection campaigns show that CNN's code and parameters stored in Flash memory are more susceptible to soft errors than related data in RAM memory. While boosting the performance, SIMD instructions supported by Arm Cortex-M4 and M7 processors increase the memory footprint and the CNN susceptibility to soft errors. In regard to the precision bitwidth variation on MobileNet CNN, the occurrence of soft errors might have a greater impact on weights, with an increase in the occurrence of critical errors when increasing the precision of activations. Finally, this confirms *the third hypothesis of the Book*, that the reduced precision optimizations impacts not only in the CNN execution, but also in the CNN code, parameters, and data stored in memory.

Although many works in the literature address dedicated to ultra-low-power integrated circuits (e.g., Arm Cortex-M family), an alternative to achieve better performance requirements is the execution of underlying ML models in more powerful processors (e.g., Arm Cortex-A family). As an effort to cover the constraints of this approach, the late results of this book comprises the assessment of the soft error reliability of a CIFAR-10 trained CNN model, which was developed based on the CMSIS-NN kernel to support multi-threaded execution. The proposed extension showed that threaded parallelism could increase the performance of the adopted CNN application with a low memory footprint overhead (i.e., less than 1%). Furthermore, the results showed that the evaluated CNN application is highly susceptible to failures as it presents critical faults in both configurations. In turn, multi-threaded versions positively impact CNN reliability while increasing the number of correct outputs, performance and, consequently, the MWTF of the adopted CNN.

Future works will focus on two main directions. The first direction comprises the soft error assessment of accelerator-based IoT devices, which are more suitable for highly computing-intensive AI applications. The second direction relies on explore the faults occurring on memory sections. Although protection mechanisms implemented in Flash and RAM memories can be employed to prevent data loss, identifying the most vulnerable CNN's code, parameters or data gives the opportunity to promote bespoke software-based solutions such as the replication of a specific function or set of parameters.

Printed in the United States
by Baker & Taylor Publisher Services